Armin Isenmann

Making new chemical elements

AF319308

Armin Isenmann

Making new chemical elements

From the beginning of the universe to the artificial elements made today

ScienciaScripts

Imprint

Any brand names and product names mentioned in this book are subject to trademark, brand or patent protection and are trademarks or registered trademarks of their respective holders. The use of brand names, product names, common names, trade names, product descriptions etc. even without a particular marking in this work is in no way to be construed to mean that such names may be regarded as unrestricted in respect of trademark and brand protection legislation and could thus be used by anyone.

Cover image: www.ingimage.com

This book is a translation from the original published under ISBN 978-620-6-76052-8.

Publisher:
Sciencia Scripts
is a trademark of
Dodo Books Indian Ocean Ltd. and OmniScriptum S.R.L publishing group

120 High Road, East Finchley, London, N2 9ED, United Kingdom
Str. Armeneasca 28/1, office 1, Chisinau MD-2012, Republic of Moldova, Europe
Printed at: see last page
ISBN: 978-620-7-74626-2

Copyright © Armin Isenmann
Copyright © 2024 Dodo Books Indian Ocean Ltd. and OmniScriptum S.R.L publishing group

Making new chemical elements

From the beginning of the universe to

the artificial elements made today

Armin Franz Isenmann

1

Contents

3

Welcome to the world of the core

Everyone is familiar with the periodic table of the chemical elements (TPE) and also knows most of the approximately 100 elements in it by name and their universal symbol. It's still common knowledge that the elements around us come in very different abundances - some elements are so common that we don't even know about them anymore (among them, oxygen, silicon, nitrogen, carbon and hydrogen), while others are rare and, among other characteristics, stand out due to their high price (hence we think of the noble metals in the first place).

But honestly: who knows why this uneven distribution? Or why the number of natural chemical elements is limited to 81[1] ? This text will hopefully provide some answers.

One more curiosity: the content of this text is not part of classical chemistry! Even so, it is currently part of the high school chemistry syllabus for the 3rd year, where the principles of radioactivity and reactions in the nucleus will be presented. However, chemistry as everyone knows it is not concerned with transformations in the nucleus of the atom, but only on the periphery where the electrons are located. And chemistry restricts itself even further: not all the electrons that are involved in reactions, but only the most energetic, outermost ones in each atom are about to make new connections with the outer electrons of other elements. They are therefore called "valence electrons".

Dear reader, if you are a chemist, let me warn you. Some of the concepts you master and apply in your day-to-day work - forget it! Here we enter another world, from different points of view:

> The bonds that are relevant to the chemist are divided into "weak bonds" and "strong bonds". The first category includes the attractions between molecules, which can be of a Van der Waals nature (1 to 3 $kJ \cdot mol^{-1}$), attractions between permanent dipoles (10 to 20 $kJ \cdot mol^{-1}$) and their special case, the hydrogen bridge (15 to 35 $kJ \cdot mol^{-1}$); these weak forces primarily explain the physical state of the material, be it solid, liquid or gas. The second category covers covalent, ionic and metallic bonds, all in the stability range of 100 to 800 $kJ \cdot mol^{-1}$. These forces describe cohesion within a molecule, between ions (salt) or between metal atoms. These "strong bonds" primarily describe the chemical reactivity of the substance.

---

[1] There are elements with atomic numbers ranging from 1 for hydrogen to 83 for bismuth, while elements 43 (technetium) and 61 (promethium) are missing. Compare p. 57.

Here we have nuclear reactions, where the cohesive forces are about 10^9 times stronger than the chemical bonds!

➢ Also related to thermodynamics are the heats involved in a physical transformation or chemical reaction. The heat involved is in the order of

20 to 40 kJ· mol⁻¹ in the case of fusion or evaporation;
100 kJ· mol⁻¹ in common chemical reactions;
400 kJ· mol⁻¹ in combustions (= complete oxidation with triplet oxygen).

Here we have nuclear reactions that release (or consume) up to 2 billion times more energy, usually indicated in MeV = Mega Electronvolts (conversion to kJ· mol⁻¹ , see footnote). 22).

➢ The physical state of reactants and products is of paramount importance in thermodynamic and kinetic considerations of chemical reactions. Not so in nuclear reactions. Even stranger: not even the chemical environment is relevant to the reactivity of a nucleus! So it doesn't matter whether we consider a nuclear reaction with beryllium hydride, BeH_2 , beryllium oxide, BeO, pure beryllium or in a metallic alloy; the speed and energy of the reaction with the beryllium nucleus are not influenced by the way this element is presented.

➢ Since the energies involved in nuclear reactions are so different from the energies involved in chemical reactions, we can say that the strong dependencies of chemical reactions on temperature - both in thermodynamics (*Van t'Hoff*) and kinetics (*Arrhenius*) - have no correspondents in nuclear reactions. The temperature governing a radioactive mineral or a nuclear reactor is of no importance. Therefore, the speed of a radioactive decay is the same whether it is observed at room temperature, inside a blast furnace or in liquid nitrogen - temperature really has no significant influence.

➢ Electrons (with the exception of the nuclear K-capture mechanism; p. 95) have no influence on nuclear reactions; often the charges are so irrelevant to the overall happening of the nuclear reaction that they can be omitted from the equations (see comment on p. 29).

➢ The speeds that are of interest to chemists are mainly those of gases. At room temperature, gases move at a speed of around 450 m s⁻¹ . Particles emitted from nuclei (called α, β and neutrons) generally have much higher kinetic energies and speeds, i.e. 10,000 to 50,000 km s⁻¹ . Finally, γ−ραψσ, a form of electromagnetic radiation, propagate at the speed of light (3· 10⁸ Km s).⁻¹

> And perhaps the strangest at the end of this list. As chemists know, mass is neither lost nor created during a chemical reaction. What gives chemistry students some problems, but which the chemist has mastered perfectly, is balancing a reaction. Here, both sides of the equation must add up to the same masses (and the same charges). This is not so true of nuclear reactions. We have to abandon the concept of mass preservation, because according to *Einstein*'s theory there is an equivalence of mass and energy (p. 64). Therefore, in reactions that are very exothermic, for example, mass loss is expected. It is therefore wise to correct the concept of the preservation of mass and believe in the preservation of energy.

However, dear chemist, not all the concepts you have mastered are for nothing or need to be abandoned.

> Reactive shocks, where we have successfully applied the concept of gas kinetics, can also be applied to nuclear reactions. Remember that the nucleus is about 10,000 times smaller than the entire atom. From this we can estimate the probability of a reactive collision between a nucleus and a particle even smaller than it (particle □□□□, neutron).
> The concept of activation energy can also be applied to nuclear considerations, where it allows conclusions to be drawn about the viability of a reaction or a process - especially a chain process (p. 114).
> Energy balances are based on the stabilities of reactants and products. This applies to both chemical and nuclear reactions.
> The first-order law of chemical kinetics is valid for describing the life of an unstable nucleus (p. 48). We can calculate the concentration of a reactive species for any reaction time.

The elements heavier than bismuth are, in ascending order of atomic mass: Po, At, Rn, Fr, Ra, Ac, Th, Pa, U, Np, Pu, Am, Cm, Bk, Cf, Es, Mf, Md, No, Lr, Rf, Db, Sg, Bh, Hs, Mt, Ds, Rg and others that are even heavier. As almost the entire mass of the atom is concentrated in the nucleus, we can conclude that these elements contain, in addition to > 83 protons, an even higher number of neutrons, which ultimately makes them unstable. Sooner or later, these nuclei decay to form lighter elements. The phenomena surrounding this decay are called "radioactivity", since the pioneering work in this area involved the element radium, Ra. This decay occurs precisely in the nucleus of the atom and not in its electronic sphere. And this has little to do with classical chemistry. As these are transformations in other chemical elements (the old dream of the alchemists who always wanted to produce gold), we can write these reactions as

A → B or A + B → C + D,

while chemical transformations, as everyone knows them, react according to the schemes

A + B → AB or AB + CD → AC + BD,

where all the elements, represented here by individual letters, are preserved.

Throughout this text we will shed light on the following subjects[2] :

> What is the origin of the light elements?
> What are the radioactive heavy elements?
> This defines the stability of nuclei and ultimately influences the abundance of elements in the universe;
> What phenomena accompany radioactivity, in terms of electromagnetic radiation, speed and energy.

Everyone expects, of course, an explanation of the most violent event made by human hands: the atomic bomb. Of course, we explain its principles, but at the very end and with emphasis on its inadmissibility.

---

[2] Warning: this text does not deal with the magnetic properties of the core.

The beginning of everything and the creation of the light chemical elements

The Universe is an incredibly differentiated and extremely complex system in terms of its structures. This is basically due to the existence of the different chemical elements. How matter was formed and how the elements were - and still are - created has always been the subject of fundamental research in the exact sciences. Among today's theoretical physicists, the *Big Bang* theory, coupled with an expanding Universe, is quite widespread, although not unanimously so.

According to today's knowledge, the Universe is 13.7 billion years old, with a margin of error of 0.2 billion more or less (the Sun, like the Earth, is estimated to be 6 billion years old). To arrive at this figure, scientists debated for almost 80 years (see Table 1). The basis of these calculations is *Hubble*'s Law:

$$v = H_0 \cdot d$$

Where:

vSpeed in km s^{-1} ;

dDistance in Megaparsecs, Mpc

(astronomical unit; 1 pc = 0.3066 light-year = 30.857$\cdot$ 10^{12} km);

H_0 *Hubble* parameter in km s Mpc$^{-1,-1}$.

Table 1 Summary of discoveries and developments in the search for the age of the Universe.

1929	American astronomer *Edwin Hubble* noticed that galaxies were moving away from each other, while the speed of the distance increased. The redshift of starlight led to the abandonment of the static model of the Universe and the origin of the *Big Bang* thesis. He estimated its age at 2 billion years.
1952	German astronomer *Walter Baade* proved that the Universe was at least twice as old as the Earth. In the years that followed, most scientists settled on 20 billion years.

1960s   The discovery of Quasars casts doubt on the *Big Bang* theory and the accelerated expansion of the Universe (*H.C. Arp; Maarten Schmidt, H. Alfvén,...*).

1980s   Construction of space telescopes. The age of the Universe is estimated at between 10 and 20 billion years.

1996   The *Hubble* Space Telescope provides data leading to an age of 8 billion.

1990s   The European satellite *Hipparcos has* measured the distance of thousands of stars with 100 times greater precision than had previously been possible. As a result, the age of the oldest stars was estimated at between 13 and 16 billion.

2000s   Dozens of researchers are redoing the calculations based on the data provided by *Hubble*.

2020   *Cosmological* monitoring of gravitational lenses via *Cosmograil*. Correction of the Hubble constant to 73 km s Mpc$^{-1 \cdot -1}$ and the age of the Universe is fixed, with a reasonable degree of certainty, at 13.7 billion years.

Throughout the history of the Universe, however, we can identify two very different eras when it comes to the creation of the elements. The first, just after the *Big Bang,* lasted only a few minutes. In this short period of time, the elementary forms of matter appeared and, finally, the nuclei of the lightest elements, hydrogen and helium. Over the next 200 million years or so, the development of the elements stagnated. Then, when the first stars appeared and illuminated the Universe, the second epoch began with the synthesis of elements heavier than helium. A Figure 1 shows the line of this development (arbitrary scale).

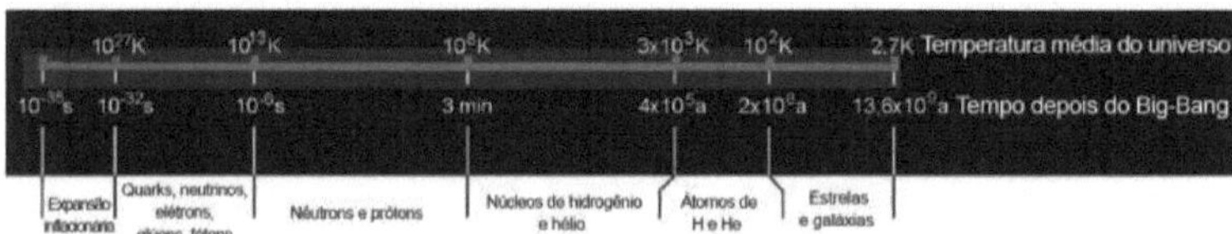

Figure 1 Timeline of cosmic events. As the Universe expanded and gradually became colder, the initial mixture of quarks, gluons, neutrinos and electrons initially formed neutrons and protons, then helium nuclei and finally the atoms of the lighter elements, hydrogen and helium. Fusion

for the heavier elements only began with the appearance of the first stars, some 200 million years later.

The atom is made up of elementary particles

Roughly speaking, the neutral atom has a diameter of approximately 1 to 2 Ångstrom (0.1 to 0.2 nm), almost all of which is occupied, under normal conditions, by the sphere of electrons (and⁻). In the center is the nucleus, in turn made up of protons (in this book represented by the symbols p, $_1^1H$ or $_1^1p$) and neutrons (represented here by the symbols n or $_0^1n$). The diameter of the nucleus is only $\frac{1}{10.000}$ of the diameter of the entire atom, i.e. its volume corresponds to less than 10^{-12} times (one trillionth) of the total volume. At the same time, it provides almost all of the atomic mass: 99.95 to 99.98%. To give you an example: a volume of 1000 m³ of iron (that would be a block of 10· 10· 10 m) contains just one cubic millimeter of nuclei, whose total mass is ~8000 tons! The rest of space is almost massless - just filled with force fields.

All the chemical qualities we know about the elements, however, are determined by the electron layer - more specifically, by the electrons in the outermost layers ("valence electrons"). The number of electrons, in turn, is determined by the number of protons in the nucleus, since each proton contributes a positive charge and an electron a negative charge. The whole leads to a neutral atom - what you would expect from a pure element in the periodic table. Finally, we identify the number of protons contained in the nucleus as being responsible for the type of chemical element.

The focus of this work will then be the creation of the atomic nucleus, an accumulation of masses whose density is around $2 \cdot 10^{14}$ g· cm⁻³, i.e. 200 million tons per cubic centimeter. The density of nuclei does not vary much when comparing elements. With a good approximation, we can estimate their diameter using the relationship

$$d = 2{,}8 \cdot 10^{-15} \cdot \sqrt[3]{m},$$

where *m is* the mass of the nucleus - ranging from 1 to 238 u for the stable elements[3] . This formula makes it possible to calculate diameters from 2.8· 10⁻¹⁵ m for hydrogen to 17.4· 10^{-15} m for uranium²³⁸ U (the average for all elements $\approx 10^{-16}$ m). Although the density is almost constant, the stability of the nuclei varies between the elements. We find the greatest nuclear stability in the

---

[3] An atomic mass *u is* now defined as 1/12 of the mass of a mole (= 6.022· 10^{23} atoms) of the isotope C.12

elements in the middle of the periodic table, i.e. near Fe, Co and Ni (Figure 14). This means that lighter nuclei tend to undergo **fusion** (p. 106) to form heavier nuclei, while nuclei of greater mass tend to **fission** (p. 101), breaking up into lighter fragments.

This is the result of exact measurements of the masses of different nuclei, as well as measuring the energy released during a nuclear reaction. It turns out that in exothermic reactions the mass of the products is slightly less than the sum of the mass of the reactants. This led to the successful application of the famous relation of *Einstein* (1905) to obtain the correspondence between energy and the mass defect:

$$E = m \cdot c^2,$$

Where E is the energy (usually indicated in Mega Electron Volts; MeV), m is the mass in grams and c is the speed of light in vacuum. With this formula we have the equivalence of

$$1 \; eV \equiv 1{,}0735 \cdot 10^{-12} kg.$$

Furthermore, with the relation $E = h \cdot v = h \cdot c \cdot \tilde{v} = e \cdot U$ you get an expression that allows you to transform energy (heat) into mass (matter), into potentials U (ionization) or into wave numbers $\tilde{v}$ (light; spectroscopy) - mutual transformations that have all been carried out in the experiment.

1kJ	$\approx$	10^{-14} kg	$\approx$	0,01 V	$\approx$	100 cm^{-1}
Calor		Massa		Potencial		Nº de ondas

Nuclear reactions can be triggered by bombarding a nucleus with particles at very high speed in particle accelerators (there is equipment that delivers up to 300 GeV!). The three most famous are the accelerator CERN (Conseil Européen pour La Recherce Nucléaire) in Geneva (Switzerland), the Lawrence Radiation Laboratory in Berkley (USA) and the proton synchrotron in Sepuchow (Russia).

These research centers have also found many elementary particles, in addition to those already mentioned *and* , $_1^1p$ e $_0^1n$. A Table 2 gives an overview of the most important particles - out of a total of > 200. It's interesting to note that for every type of elementary particle there is an antiparticle; a few are even particle and antiparticle at the same time. The science behind these is a branch of theoretical physics known as **quantum chromodynamics** (QCD).

Table 2 The most important elementary particles that make up solid matter. Note that the anti-particles are represented by a bar above the symbol.

Leptons (spin ½)					Quarks (spin ½) (they don't exist in isolation)			
Name	**Symbol**	**Mass (MeV)**	$\tau_{1/2}$ (s)	**Load**	**Name**	**Symbol** $q, \bar{q}$	**Mass (MeV)**	**Load** $q, \bar{q}$
Neutrinos	$\nu_e, \bar{\nu}_e$	≈ 0	stable	0, 0	Up	$u, \bar{u}$	300	+2/3 -2/3
	$\nu_\mu, \bar{\nu}_\mu$	≈ 0	stable	0, 0	Down	$d, \bar{d}$	300	-1/3 +1/3
	$\nu_\tau, \bar{\nu}_\tau$	≈ 0	stable	0, 0	Strange	$s, \bar{s}$	45	-1/3 +1/3
Electron	e^-, e^+	0,5	stable	-1, +1	Charmed	$c, \bar{c}$	1500	+2/3 -2/3
Muon	μ^-, μ^+	106	$\approx 10^{-6}$	-1, +1	Bottom	$b, \bar{b}$	4900	-1/3 +1/3
Tauon	τ^-, τ^+	1800		-1, +1	Top	$t, \bar{t}$	>18.000	+2/3 -2/3

Hadrons (pairs or trios of quarks; products of strong interactions)

Months (spin 0)					Baryons (spin ½)				
Name	**Symbol**	**Mass (MeV)**	$\tau_{1/2}$	**Quark str.**	**Name**	**Symbol**	**Mass (MeV)**	$\tau_{1/2}$	**Quark str.**
Pions	π^0	135	$<10^{-16}$	$\bar{u}u/\bar{d}d$ 50:50	Nucleons:				
	$\pi^+ \ \pi^-$	140	$\approx 10^{-8}$	$\bar{d}u d\bar{u}$	Proton	$p^+ \ p^-$	938	stable	uud $\bar{u}\bar{u}\bar{d}$
Káons	$k^+ \ k^-$	494	$\approx 10^{-8}$	$\bar{s}u$	Neutron	$n \ \bar{n}$	940	$\approx 10^3$	udd $\bar{u}\bar{d}\bar{d}$

					Hyperons:				
	$k^0 \ \bar{k}^0$	498	$\approx 10^{-10}$	$\bar{s}d$					
M σον η	η^0	549	$\approx 10^{-19}$	$\bar{u}u$ /$\bar{d}d$ /$\bar{s}s$		Λ^0 $\bar{\Lambda}^0$	1116	$\approx 10^{-10}$	uds $\bar{u}\bar{d}\bar{s}$
Charming months	D^0 $\bar{D}^0$	1863		$\bar{u}c$		Σ^+ $\bar{\Sigma}^+$	1189	$\approx 10^{-10}$	uus $\bar{u}\bar{u}\bar{s}$
	$D^+ \ D^-$	1868		$\bar{d}c$		Σ^0 $\bar{\Sigma}^0$	1192	$\approx 10^{-19}$	uds $\bar{u}\bar{d}\bar{s}$
	η_c^0	2980		$\bar{c}c$		Σ^- $\bar{\Sigma}^-$	1197	$\approx 10^{-10}$	dds $\bar{d}\bar{d}\bar{s}$
	B^- $\bar{B}^+$	5260		$\bar{u}b$		Ξ^0 $\bar{\Xi}^0$	1315	$\approx 10^{-10}$	uss $\bar{u}\bar{s}\bar{s}$
	$B^0 \ \bar{B}^0$	5260		$\bar{d}b$		Ξ^- $\bar{\Xi}^-$	1321	$\approx 10^{-10}$	dss $\bar{d}\bar{s}\bar{s}$
	y^0	9460		$\bar{b}b$		Λ_c^+ $\bar{\Lambda}_c^+$	2273		udc $\bar{u}\bar{d}\bar{c}$

Without going into detail, we'll just mention some of the principles of QCD:

- Quarks are never found alone, but always in pairs $\bar{q}q$ to form mesons or in trios qqq or $\bar{q}\bar{q}\bar{q}$ to form baryons. Outside their typical distance of 10^{-15} m from each other - for less or more - the quarks suffer an increase in potential energy of the order of a few GeV. With such a deep potential well, the forces that exist between the quarks really deserve the name "strong interaction". These attractions are even stronger than those that exist in nuclei, between protons and neutrons.
- If a particle collides with its antiparticle, the masses can be extinguished and transformed into a large amount of energy, emitted in the form of rays ☐☐(" annihilation ").
- It is the set of quarks in pairs or trios that we should call strong atomic interactions - and not the set of electrons and nuclei, nor the set of protons and neutrons which, in turn, is already a subsequent interaction between baryons. This being the case, it would also be more appropriate to preserve the expression "elementary particle" for the pairs and trios of

quarks and not for atoms (set of nuclei and electrons) or nuclei (set of protons and neutrons).

> It is possible to transform matter into energy and vice versa. The relationship between electronvolts and kilograms given above applies, or for larger quantities:

1 million faradayvolts = 96,485 million electronvolts = 1.0735 milligrams.

The etymology of elementary particles helps us to visualize their main properties:

- Leptons - leptos (Greek) = light;
- Hadrons - hadros (Greek) = fat, robust, strong;
- Mesons - meson (Greek) = in between;
- Baryons - barys (Greek) = heavy;
- Hyperons - hyper (Greek) = above.

With the exception of the mesons, all these particles are fermions, with spin ½. In addition to these families of elementary particles, shown in Table 2we still have to count how many physical particles cause interactions; they have no mass of their own and are Bosons (with spin 0, 1 or 2):

- Gluons (from the English: *to glue*; responsible for the strong interactions between quarks;
- Photons - photos (Greek = light); responsible for the electromagnetic interaction between two charged particles;
- W or Z bosons; responsible for weak interactions (such as chemical bonds);
- (Grávitons (Latin = gravitas = heavy); responsible for gravitational attraction).

With today's knowledge, backed up by experiments, we can build a standard model of elementary particle physics (Figure 2), which includes:

> Six types of quarks, each with three different colored charges;
> Six types of leptons, namely three charged leptons and three neutrinos;
> Twelve types of gauge bosons, namely:
> - a photon for electromagnetic interaction;
> - eight gluons for the strong interaction;
> - three bosons (Z^0, W^+ and W^-) for the weak interaction; and finally,
> The *Higgs* boson.

Modelo Padrão das Partículas Elementares

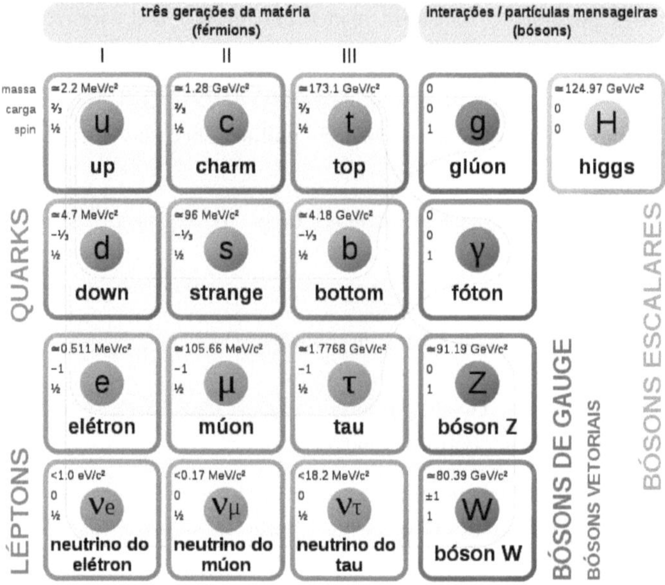

Figure 2 The Elementary Particle Model with the bosons in the last column.

Source: wikimedia commons (free image): https://pt.wikipedia.org/wiki/Modelo_Padrão.

The story shows its face for the first time

The Universe originated in a huge hot explosion, the "*Big Bang*" - this is widely accepted today. However, what happened immediately after the *Big Bang,* in the first fractions of the first second, is still pure speculation. Only after the exponential expansion, a phase in which the Universe probably expanded by a factor of 10^{29} , do theories describe the situation a little more accurately. Thus, 10^{-33} seconds after detonation, the Universe consisted of a soup of particles and antiparticles, a mixture of heavy X bosons, quarks, electrons and gluons whose temperature was approximately 10^{27} Kelvin.

Shortly afterwards, the heavy X bosons disintegrated into quarks and their antiparticles. The condition for such elementary reactions was a drastic drop in temperature and energy density, which became lower than the values needed to preserve these massive particles. And this change of state was caused by the rapid (adiabatic) expansion of the Universe. This process should have provided

16

exactly the same number of quarks and antiquarks. However, there was a slight excess of one normal quark for every 10 billion quark/antiquark pairs. Soon afterwards, when the energy decay of the pairs began, quarks and antiquarks annihilated each other, leaving only the quarks that hadn't found antiparticles. It is thanks to this asymmetry in the decay of the X bosons that there is something with mass left in our Universe. A completely symmetrical decay would have destroyed all matter and turned it into radiation ☐, so the Universe would have remained structureless, without galaxies, stars and planets.

Approximately one millionth of a second after the *Big Bang*, at a temperature of around 10 trillion Kelvin, the next step in the formation of matter took place. At this point, the energy density had dropped so much that the remaining quarks could no longer exist as independent particles. A neutron was formed from an *Up* quark and two *Down* quarks; a proton was formed from a *Down* quark and two *Up* quarks. Thus, the atomic nuclei of the lightest element, hydrogen, filled the Universe in just one millionth of a second after the *Big Bang*.

Around a second later, the Universe was transformed into a gigantic fusion reactor. Over the next three to four minutes, in the **primordial nucleosynthesis** phase, a neutron and a proton fused together to form the nucleus of deuterium (atomic symbol: ^2_1H or D), at a temperature of around one billion Kelvin. When two deuterium nuclei collided, tritium (atomic symbol: ^3_1H He or T), an isotope of hydrogen with two neutrons in the nucleus, or else helium-3 (atomic symbol:[3] He or ^3_2He), an isotope of helium made up of two protons and one neutron. Finally, a deuterium nucleus fused with a[3] He nucleus or a deuterium nucleus with a tritium nucleus. Both reactions led to the appearance of[4] He (also known as the α particle, one of the rare heavy Bosons), under the emission of a proton or neutron respectively:

$$^2_1H \; + \; ^3_2He \; \xrightarrow[-\,p]{} \; ^4_2He; \qquad\qquad ^2_1H \; + \; ^3_1H \; \xrightarrow[-\,n]{} \; ^4_2He.$$

The union of two deuterium nuclei, the capture of a proton by a tritium nucleus or the union of a neutron and a helium-3 nucleus are other reactions that led to the helium-4 nucleus:

$$^2_1H \; + \; ^2_1H \; \longrightarrow \; ^4_2He; \qquad ^3_1H \; + \; p \; \longrightarrow \; ^4_2He; \qquad ^3_2He \; + \; n \; \longrightarrow \; ^4_2He.$$

Finally, some tritium and helium nuclei were combined to form the nucleus of the isotope of ^7_3Li.

In these processes, almost all the neutrons were consumed to build the nuclei of the first chemical elements. After the completion of primordial nucleosynthesis, the matter in the Universe, expressed as a percentage by weight, consisted of approximately 75% hydrogen, 24% helium-4, 0.001% helium-3, 0.002% deuterium and 0.00000001% lithium-7.

However, it took around 400,000 years for atomic nuclei to become complete atoms. Only when the temperature of the Universe dropped to around 3000 Kelvin did the photons no longer have enough energy to ionize the atoms. Now, the atomic nuclei could capture electrons and bind them permanently in their orbits. At the end of these recombination processes, the Universe consisted essentially of atomic hydrogen gas and helium, a mixture distributed almost homogeneously in space. Note that at this time there was still no sign of the other elements in the periodic table.

Light at last - the rise of the stars

After this fulminant beginning, the development of the elements stagnated. But the Universe continued to expand and cool. As a result, there were soon no more photons whose wavelength corresponded to visible light. At that time, for a hypothetical observer, everything would have been absolutely dark, without a single source of light. This **"dark age of the Universe"** lasted approximately 200 million years. During this time, gravity (which is the weakest cohesive force of all known forces) pulled in more and more gas which agglomerated into huge clouds. Eventually, gigantic balls of gas made of hydrogen and helium formed, in which the pressure and temperature increased continuously. At a pressure of around 200 billion atmospheres and a temperature of around 15 million Kelvin, the hydrogen protons collided with such force that they were able to cross the repulsive Coulomb barrier, a type of tunneling, and merge to form more helium. At that moment, around 200 million years after the *Big Bang*, the first stars began to shine.

These primordial stars, also called **"population III stars"**, differed in many ways from the current generation of stars (population I): they consisted exclusively of hydrogen and helium, had masses ranging from a few hundred to 1,000 solar masses and their surface temperatures of around 100,000 Kelvin gave them a luminosity millions of times greater than that of the Sun. They burned their fuel so quickly that their lifespan was limited to a few million years. But inside they generated the first heavy elements and, as they died in huge supernova explosions, they threw them into the gas of the surrounding cloud. The next generation of stars that formed from these clouds was therefore already enriched with elements other than helium, with "metals", as astronomers call chemical elements heavier than helium.

Today, most newly formed stars are less massive than the Sun and live for tens of billions of years. How can this be explained? In atomic clouds, some elements tend to combine to form simple molecules, for example carbon dioxide. These molecules cool the gas in the cloud, because during their collisions they absorb the energy of the collision and transform it into vibrations and rotations. These forms of energy, in turn, can be dissipated in the form of electromagnetic radiation in the infrared and radio regions. The radiation leaves the cloud almost unhindered, so that the cloud's temperature drops to a few degrees above absolute zero. "Low temperature" is synonymous with low gas pressure. Under this condition, gravitation is able to compress small cloud fragments, driving against the internal pressure of the gases, forming new stars. In the original clouds, which were still free of metals, only the hydrogen molecules contributed to the cooling process. It was therefore not possible to reach temperatures below 150 to 100 Kelvin. As a result, large masses of gas had to come together in order for gravity to gain the upper hand over the cloud's internal pressure.

Within stars, regardless of which population they belong to, the processes that lead to the formation of elements obey the same rules - despite slight variations due to different metallicities. We can therefore describe the events in a generalized way:

Star - the factory of the elements

The fusion processes that produce the chemical elements take place inside stars. Hydrogen serves as the starting material. When the temperature in the center of the star rises (due to gravitational compression) to around 15 million Kelvin, two protons fuse to form a deuterium nucleus, under the emission of a positron and a neutrino:

$$2\,{}^{1}_{1}H \longrightarrow {}^{2}_{1}H + e^{+} + \nu$$

By attaching another proton, a helium-3 nucleus is formed and, finally, two helium-3 nuclei fuse to form helium-4. The excess of two protons is available for a new proton-proton reaction chain (Figure 3a). In short, this "hydrogen burning" consumes four protons to build a helium nucleus, which is why we call it the "**pp cycle**" (*Critchfield*, 1938):

$$2\,x\ |\ {}^{1}_{1}H + {}^{1}_{1}H \longrightarrow {}^{2}_{1}H + e^{+} + \nu_{e} + 1{,}44\,\text{MeV} \quad 2\,x\ |\ {}^{1}_{1}H + {}^{2}_{1}H \longrightarrow {}^{3}_{2}He + 5{,}49\,\text{MeV} \quad {}^{3}_{2}He + {}^{3}_{2}He \longrightarrow {}^{4}_{2}He + 2\,{}^{1}_{1}H + 12{,}85\,\text{MeV}$$

$$4\,{}^{1}_{1}H \quad\longrightarrow\quad {}^{4}_{2}He + 2\,e^{+} + 2\,\nu_{e} + 26{,}71\text{ MeV}$$

At even higher temperatures and in stars with a sufficient proportion of carbon and oxygen to act as "catalysts", the "**CNO cycle**" takes precedence (*Bethe, von Weizsäcker,* 1937). Here too, four protons are consumed to build a helium-4 nucleus. But the path is different, because carbon and oxygen act as catalysts and nitrogen represents the intermediate product (Figure 3b). The star owes its longevity to the low concentrations of C and O in its interior:

$$ {}^{1}_{1}H + {}^{12}_{6}C \longrightarrow {}^{13}_{7}N \qquad + 1{,}95\text{ MeV}$$

$$ {}^{13}_{7}N \longrightarrow {}^{13}_{6}C + e^{+} + \nu_{e} + 2{,}22\text{ MeV}$$

$$ {}^{1}_{1}H + {}^{13}_{6}C \longrightarrow {}^{14}_{7}N \qquad + 7{,}54\text{ MeV}$$

$$ {}^{1}_{1}H + {}^{14}_{7}N \longrightarrow {}^{15}_{8}O \qquad + 7{,}35\text{ MeV}$$

$$ {}^{15}_{8}O \longrightarrow {}^{15}_{7}N + e^{+} + \nu_{e} + 2{,}71\text{ MeV}$$

$$ {}^{1}_{1}H + {}^{15}_{7}N \longrightarrow {}^{4}_{2}He + {}^{12}_{6}C + 4{,}96\text{ MeV}$$

$$4\,{}^{1}_{1}H \quad\longrightarrow\quad {}^{4}_{2}He + 2\,e^{+} + 2\,\nu_{e} + 26{,}73\text{ MeV}$$

Note that the energy released during nuclear **fusion** processes is an order of magnitude greater than the energy involved in the **fissions** that take place in today's nuclear power plants, when related to the mass of starting material.

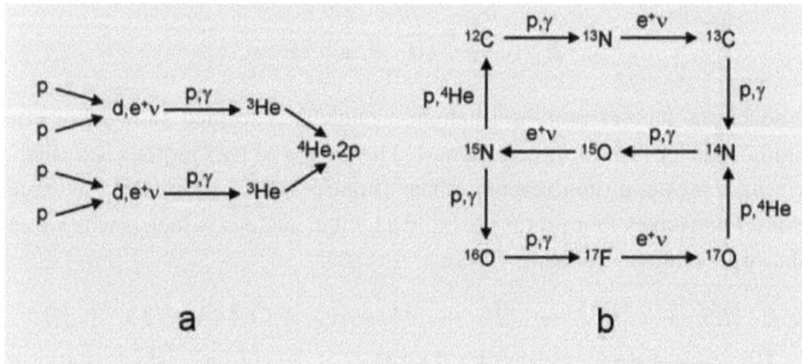

Figure 3 The fusion of hydrogen to helium occurs in stars via the pp cycle (left) or the CNO cycle (right). At temperatures below 18 million

Kelvin, the pp cycle dominates; above this temperature, the CNO cycle prevails.

Stars the size of the Sun "burn" their hydrogen reserves in about 10 billion years. During this period they emit energy, mainly in the form of positrons and neutrinos, as described in the pp and CNO cycles. Here are the impressive data for the Sun, which is a slightly larger than average star in the Milky Way (see further details about the Sun, the Earth and the Universe on p. 147). 144):

> The Sun's radiation power: $3.72 \cdot 10^{23}$ kW (1 W = 1 kJ s^{-1}), this corresponds to 61,300 kW for every square meter of the Sun's surface, estimating it at $6.072 \cdot 10^{18}$ m^2.
> Energy production requires around 580 million tons of hydrogen every second. But this isn't worrying, as the Sun's hydrogen reserves are still 10^{27} tons. Given that in a billion years (the age of the Sun is estimated at 6 billion) only 1/50th of the H reserves are burned, we can still bet on a long life for our parent star.
> Due to radiation, the Sun's mass decreases (*Einstein*'s equation on p. 12) by about 4.1 million tons every second. Over a period of one billion years this corresponds to a loss of $1.3 \cdot 10^{17}$ t, i.e. 10^{-7} times its total mass.
> The temperature at the center of the Sun is around $15 \cdot 10^6$ K, while on its surface it is "only" 5500 - 5800 K.

When all the protons in the star's core are burned to helium-4, the central atomic fire goes out. Helium is then only produced in a narrow layer around the center of the star. The cooling of the center allows its density to increase, i.e. the helium core shrinks under its own gravity. Now a strange game begins: with the increase in density, the gravitational acceleration that the core exerts on the layers of gas also increases. This raises the pressure in the hydrogen-burning layer and, as a result, the fusion processes accelerate and release more and more energy. At this stage, a star with dimensions similar to our Sun inflates into a **red giant**, with a radius of up to 50 times the solar radius and a luminosity of several hundred times that of the Sun.

Eventually, the pressure and temperature in the helium nucleus reach such high values that other nuclear reactions, whose activation barriers are higher, can also begin. Now,^{4} He burns to carbon through the 3–α process and the capture of another helium nucleus creates^{16} O. Through other capture ρεαχτιονσ–α, σ©o also accumulated in small quantities the new nuclei of^{20} Ne,^{24} Mg and^{28} Si. During these processes, the star returns to some of the sun's rays. As helium burning occurs at higher temperatures than hydrogen burning, this phase is significantly shorter than hydrogen burning and takes "only" 50 million years in a star similar to our Sun.

The next steps at the end of helium burning are similar to those at the end of hydrogen burning: again, a fusion zone forms around the stellar center, which now consists essentially of carbon and oxygen. There are now two distinct burning zones in the star: in the inner layer, helium reacts to carbon and oxygen; in the outer layer, hydrogen continues to convert to helium.

And, for the same reasons that led the star to grow at the end of the hydrogen burn, it is now inflating again into a red giant, which is, however, many times larger than the first time. Stars of this size are no longer able to gravitationally retain their outer layers of gas. Therefore, strong winds develop at this stage, at speeds of up to 30 km/s. They mobilize part of the star's outer layer, containing the nuclei of various elements, in the form of a planetary fog that moves away into infinite space. We note that, unlike hydrogen and helium, which emerged as atomic nuclei just a few minutes after the *Big Bang, it* took billions of years for a star like the Sun to contribute to enriching the interstellar medium with elements heavier than helium.

Stars with less than eight solar masses lose up to 60% of their mass thanks to stellar winds. Although the star's core continues to shrink over the following years, its internal temperature no longer reaches the values required for additional nuclear fusions, due to the star's reduced mass and decreased gravitational pressure. The atomic fire is finally extinguished. What remains is a hot lump of carbon and oxygen several thousand kilometers in size, a **white dwarf** that cools over the years and loses its luminosity.

The heaviest stars go further

Stars with more than eight solar masses show a different development. The fusion processes are almost the same until the helium is burned, although they occur more quickly. For example, the burning of hydrogen in a star of 25 solar masses takes only about seven million years, and the burning of helium ends after about 700,000 years. Although these massive stars lose substance due to stellar winds, the remaining stellar mass is large enough to allow the core temperature to rise through gravitational pressure to a level where carbon fusion at ^{20} Ne, ^{23} Na and ^{24} Mg ignites. In a star of 25 solar masses, this phase lasts around 400 years.

If the carbon supply is also exhausted, the zone where the carbon is burned moves from the center to a new (third) fusion shell around the star's core.

However, the core is further compressed by gravitation and the pressure and temperature in the center increase again. The burning of neon then begins at around one billion degrees, with[20] Ne initially splitting into[16] O and a particle α, by photo-disintegration. From these fission products,[20] Ne,[24] Mg and[28] Si are built again. From two billion degrees, the burning of oxygen begins with Mg,[2428] Si, S, P and[313132] S being fusion products. Finally, at three billion degrees, part of the silicon nuclei is broken down into particles □□via photo-dissociation, which fuse with the rest of the silicon to form iron and nickel.

It only takes a day for silicon to burn inside a star of 25 solar masses. As the iron core has the greatest stability of all the elements (Figure 14), the fusion of even heavier elements could only take place under energy supply. As a result, the chain of fusions breaks in iron. At the end of this development, the star resembles an onion with a central area burning silicon, surrounded by five layers in which (from the inside out) oxygen, neon, carbon, helium and hydrogen are the fuels (Figure 4).

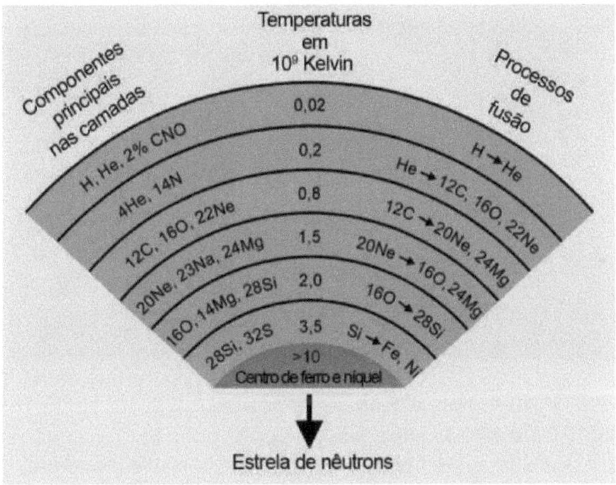

Figure 4 Structure of a 25 solar mass star. At the end of its life, the formation of a heavy star resembles that of an onion. The layers are fusion zones, gradually colder from the inside out. The products generated by the previous reaction phases burn in the inner layers, producing new nuclei with larger masses. When all the silicon finally fuses into iron, the star explodes into a type II supernova.

When the silicon runs out in the center of the superstar, the energy source runs out and gravity finally overcomes the stabilizing internal pressure. The star now collapses under its own weight in a fraction of a second. The masses of gas crash into the hard cores in the center, causing shock waves that propagate outwards along with a stream of fast neutrinos and blow the star apart in a huge explosion, a **type II supernova**. In this furious finale, all the layers are disintegrated from the core and thrown into space. As a result, the elements lighter than iron spread out radially in the path of a disk. The tremendous shock of the supernova carries the fusion products millions of light years away into the interstellar medium (Figure 5). What remains of the heavy star is just a small, super-dense neutron star that is only a few tens of kilometers in diameter. Such a stellar disaster is revealed to an observer by a luminosity that briefly exceeds the totality of all the stars in a galaxy.

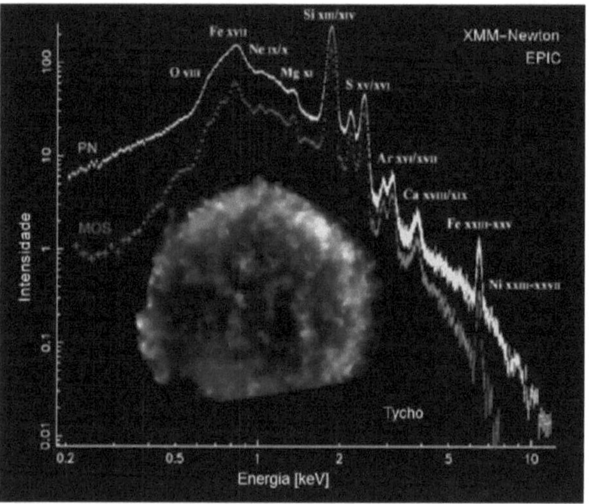

Figure 5   Remnant of the *Tycho* supernova. In 1572, the Danish astronomer *Tycho Brahe* observed a supernova in the constellation of *Casseopeia*. Since its supernova explosion, the particle cloud has expanded to a diameter of approximately 24 light-years. The gas clouds are still a few million degrees high and emit X-rays, which are received by the *Chandra* X-ray satellite and produce a false-color image. The spectrum of the explosion clouds (blue and green line) was recorded by the NASA / ESA *X-ray* satellite *XMM-Newton* and shows a large number of emission lines, in which the ionized atoms of the elements oxygen, magnesium, silicon, sulphur, argon, calcium, iron and nickel stand out.

[Photos: APOD from 12/9/2002 and http://wave.xray.mpe.mpg.de/rosat/calendar/2002/sep]

APOD   =   Astronomical   Picture   of   the   Day.   Accessible   at http://antwrp.gsfc.nasa.gov/apod/archivepix.html

Neutron by neutron towards the heavier elements

Elements heavier than iron are built in stars through the *s* and *r* processes, which are fundamentally different from the thermonuclear reactions previously considered. Both are based on reactions (n, $\square$), i.e. the capture of neutrons followed by decay $\square$. This reaction sequence is also carried out artificially in cyclotrons (CERN, for example. See p. 12). However, they differ significantly with regard to the time constant τ_n that leads to neutron capture. The "*s*" comes from "*slow*" and the "*r*" from "*rapid*", referring to the speed of neutron capture. The *s* and *r* processes contribute approximately equally to the production of heavy elements.

In the ***s-process***, a neutron is captured for the first time by a stable parent isotope, forming an isotope with a mass increased by one (Figure 6). Continuous neutron additions lead to gradually heavier isotopes until finally an unstable radioactive isotope is reached, which decays by converting a nuclear neutron into a proton, emitting an electron *and* and an antineutrino. $\bar{\nu}_e$. Thus the next higher element is ready. From this element, heavier and heavier stable isotopes are built up by capturing neutrons until the next unstable isotope is reached. In this way, heavier elements are gradually built up by neutron capture and subsequent decay β (more detailed explanations on p. 85). 83).

The process takes place mainly in the helium-burning areas of pulsating stars, red giants. At the prevailing temperatures of around $3 \cdot 10^8$ Kelvin the following reactions take place:

^{22}Ne $(\square, n)^{25}$ Mg and ^{13}C $(\square, n)^{16}$ O. [4]

These reactions lead to high neutron densities of around 10^8 cm^{-3} . In this environment, the neutron capture times τ_n are of the order of several tens of years. As the half-life of the average $\square\square\square$em decay varies from a few days to a year, there is enough time to allow the next higher element to emerge from an unstable isotope via β, decay before another neutron is added. Starting with ^{56}Fe, the elements up to bismuth (^{209} Bi) are built up by the *s-process*. The process does not continue, because the implantation of more neutrons leads to unstable nuclei that disintegrate by decay $\square$.

---

[4] The annotation in brackets represents the attacking particle first and the disintegrated particle of the new nucleus second; compare p. 66.

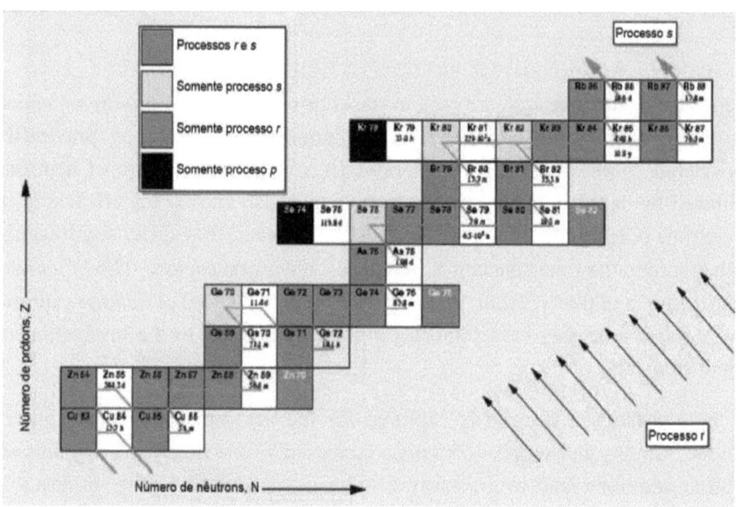

Figure 6 Elements heavier than iron are produced through the *s* and *r* processes, in which the addition of a neutron to a stable isotope and the subsequent β⁻ decay result in an isotope of the next heaviest element. Some proton-rich elements that cannot be reached by the *s* or *r* processes are created using the *p* process.

[Image: www.ifs.physik.uni-stuttgart.de/Personal/MJaeger/picture/sprocess.gif].

In addition to neutron-poor elements, which also arise through the *s-process,* neutron-rich elements such as thorium and uranium are synthesized through the **r-process**. The knowledge gained from research into the development of heavy stars suggests that the *r-process* occurs mainly behind the outgoing shockwave experienced in a type II supernova explosion, where temperatures of up to $2 \cdot 10^9$ Kelvin and neutron densities greater than 10^{20} cm^{-3} are reached over a short period of time. Possible sources of these neutrons are reactions of the type (p, n), ($\square$, n) and (n, 2n). In this environment, only milliseconds pass before a neutron connects with a nucleus, so we can say that $\tau_n \ll \tau_\beta$. As a result, several neutrons can connect to a stable atomic nucleus in a short time, without this process being interrupted by decay $\square$. This continues until an equilibrium in reactions of the type [(n, $\square$) : ($\square$, n)] is established. At the end of the addition chain, the nucleus is displaced by up to 20 mass units into the region of neutron-rich nuclei. When the flow of neutrons finally decreases, the isotopes return to their most stable level due to successive decays β (which implies the transformation of a neutron into a proton).

Apart from the elements generated through the *s* and *r* sequences, there are several nuclei that cannot be reached by neutron capture processes. Their origin can be explained by a different phenomenon, called the ***p-process, the*** details of which are not yet fully understood. It works, like the *r-process*, in the explosive layers of a supernova. According to this theory, proton-rich isotopes are formed, in the case of small-mass nuclei, exclusively through proton capture, or in the case of larger-mass nuclei, through proton capture reactions with subsequent positron decay β^+ (implying the transformation of a proton into a neutron). The parent nuclei are those previously synthesized by the *s* and *r* processes.

Is the creation of the elements over?

The creation of elements should not be considered finished, because the corresponding processes are still taking place today inside stars. For example, in the spectra of red giants, stripes have been found that prove the unstable isotope of technetium (p. 90), whose lifetime is considerably shorter than the typical development time of the star in question. In other words, if technetium wasn't continuously recreated, nothing would be found.

Over the last three decades, nuclear chemists and physicists have also managed to generate small quantities of heavy elements in the laboratory by bombarding atomic nuclei with neutrons (p. 83), protons (p. 78) or other atomic nuclei - including heavier nuclei, up to $^{20}_{10}Ne^{10+}$ (p. 87). The aim of this research is to produce elements, extremely rich in neutrons, which remain in the TPE in addition to uranium. Unfortunately, these "transuranic elements" generally disintegrate after a short period of time. But, in principle, the artificial production of a stable transuranic element should not be impossible. When it comes to producing these new elements on a large scale, however, the monopoly is still with the stars.

Natural radioactive elements

The law of Soddy, Fajans and Russel

The phenomenon of the natural decay of the nucleus is found mainly, but not exclusively, in the heavier chemical elements. Since the sequence (main order principle in TPE is the number of protons in the nucleus) we can say that a high order number of the element (> 83) represents at the same time the number of positive charges in the nucleus. An accumulation of charges obviously makes the nucleus unstable. An unstable nucleus, in turn, frees itself of particles in order to transform itself into a more stable nucleus.

The nucleus of the elements consists of protons and neutrons, in approximately equal numbers (slight excess in neutrons; see Figure 11). This means that the **atomic number Z** (= number of protons; whole number in TPE) is accompanied by the **atomic mass M** (= number of [protons + neutrons]; both of a mass of approximately 1 u; exact values later), hence the atomic mass M is approximately twice the order number Z. Interestingly, neutrons can "react" with each other, producing negatrons "e⁻ " and positrons, "e⁺ ". We assume that radioactive decay can produce protons, neutrons, any combination of these two particles, and/or electrons with a negative or positive charge. The experience of more than 100 years, however, which involves energy aspects (p. 64), restricts the number of particles emitted during natural decays (i.e. not forced by means of an accelerator) drastically. It has been shown that natural radioactivity involves only two of these multiple possibilities:

> ➤ From the unstable nucleus comes a particle made up of two protons and two neutrons - what we identify as the helium nucleus, He^{2+} , or
> ➤ Negative electrons are emitted from the unstable nucleus, and⁻ .

Popularly, these particles are called the **α particle** or helion in the first case and the **β particle** or negatro in the second case. The emission of other combinations, such as protons and neutrons, or protons, neutrons or positrons alone, is not observed in natural decay. We emphasize that this refers to natural processes; this does not mean that this atomic debris is not possible in artificial nuclear reactions (p. 69). On the contrary: the variability of the particles emitted under the delivery of immense energy densities - this is what happens in artificial processes - is vast (details on p. 95). 93).

The restriction to particles α and β emanating from natural processes already suggests that the combination of two protons and two neutrons is especially stable. Their presence seems to be fundamental in the construction of new elements - an aspect that we will return to when discussing the stability of chemical elements, including helium.

What are the consequences of emitting a nucleus $^4_2He^{2+}$? [5] After the emission of the particle α, the element has another name, it has two fewer positive charges and its mass is reduced by 4. We can say that the new element has moved two places to the left on the TPE and has a reduced atomic mass of 4 u.

For example, the metal radium, whose atomic mass is 226 u and nuclear charge 88, transforms into the noble gas radon, with respective values of 222 and 86, when it emits a particle α:

$$^{226}_{88}Ra \;\rightarrow\; ^{222}_{86}Rn^{2-} \;+\; ^4_2He^{2+} \;+\; energia \qquad \text{Equation 1}$$

For a chemist, this nuclear reaction is rather strange: a metal is gone and two noble gases appear! And one more fact, which we'll specify later: the energy of this exothermic reaction is around 10^8 times (in a word: a hundred million times!) greater than that of an "ordinary" chemical reaction.

The emission of an electron and⁻ (= particle □) results in the transformation of a neutron in the nucleus into a proton. Thus, its positive charge increases by one and it takes on the characteristics of the element to its right on the TPE. As the electron has a much lower mass than the proton or neutron (p. 143), and also due to the fact that⁻ as soon as it is emitted from the nucleus it can be trapped in the electronic sphere of the new element, we can say that the atomic mass is unchanged in this decay. An **isobar of** the decayed element **is** formed.

Example: the metal actinium with an atomic mass of 227 u (exact value: 227.09844) and a nuclear charge of 89 emits a negatron β^- and is transformed into the element thorium with the same mass of 227 u (exact: 227.09836) and a nuclear charge of 90:

$$^{227}_{89}Ac \;\rightarrow\; ^{227}_{90}Th^+ \;+\; ^{\;\;0}_{-1}e^- \;+\; energia \qquad \text{Equation 2}$$

Both here and in Equation 1, <u>the charges that accompany the products are chemically irrelevant,</u> since the positive and negative ions, immediately after decay, are neutralized by receiving or releasing an electron from the outer

---

[5] Look at the annotation: in the bottom left-hand corner the number of protons which determines the name of the element; in the top left-hand corner the rounded mass of the particle, the atomic mass; in the top right-hand corner the electric charge of the particle.

sphere of the atom, respectively. For this reason, we will leave out the notes on charges in nuclear reactions.

These characteristics of particle emissions α and β are summarized in the basic laws of radioactivity, first formulated in 1913 by *Soddy, Fajans* and *Russell* (**elementary displacement laws**).

The natural decay series

In most natural decays, a new element is formed, which in turn is radioactive, i.e. unstable. This means that the decay continues and a series of decays are evident, typical and determined by three factors:

- The initial element, the first and usually heaviest of each series;
- Decay processes (radiation α or $\square$);
- The stability of the new elements, which in the end can be very different, is a factor that determines the lifespan and, ultimately, the abundance of each element in the series, in the natural mix of all the participants.

Naturally, there are three series of heavy element decays:

- ❖ The first series, known as the **actinium series,** begins with the isotope $^{235}_{92}U$ ("Actinium uranium"; half-life[6] of $t_{1/2} = 703,800,000$ years)
- ❖ The second **series** is the **Uranium Series** and begins with the most abundant isotope of uranium, $^{238}_{92}U$ ($t_{1/2} = 4,468,000,000$ years)
- ❖ The third series is called the **Thorium Series**[7] , starting with $^{232}_{90}Th$ ($t_{1/2} = 23,420,000$ years) which in turn was created by the $^{236}_{92}U$ by decay $\square$.

All the series and the characteristics of their participants are listed in Table 3.

For those who paid attention: the masses of these starting elements in the series can be represented by (*4n* + 3) in the case of the actinium series, by (*4n* + 2) in the case of the uranium series and by (*4n* + 0) in the case of the thorium series; *n* in this context represents an integer. Where is the (*4n* + 1) series? It was later discovered to be an artificial series that was soon named the **Neptunium Series** (Figure 7). Neptunium, in turn, can be deduced from the isotope $^{233}_{92}U$ whose half-life is 159,000 years, i.e. much shorter than the beginners in the natural decay series. Therefore, this element and the isotopes that appear in this series

---

[6] In radioactive processes, the "half-life" or "semi-disintegration period", $t_{1/2}$, of a radioisotope is the time needed to disintegrate half the original mass of this isotope. A closely related index is the "average life expectancy" τ (less commonly used in the context of radioactivity). In the case of exponential decay, it is the period of time after which the size has decreased to a fraction $1/e \approx 36.8\%$ of the initial value, i.e. $\tau = {t_{1/2}}/{\ln 2}$.

[7] Or "pacifist series", as it would be explained on page 107.

have already disappeared over the course of the Earth's life. On the other hand, it is very likely that the elements in this unnatural series already existed in the past - as did all the other elements unknown in nature today.

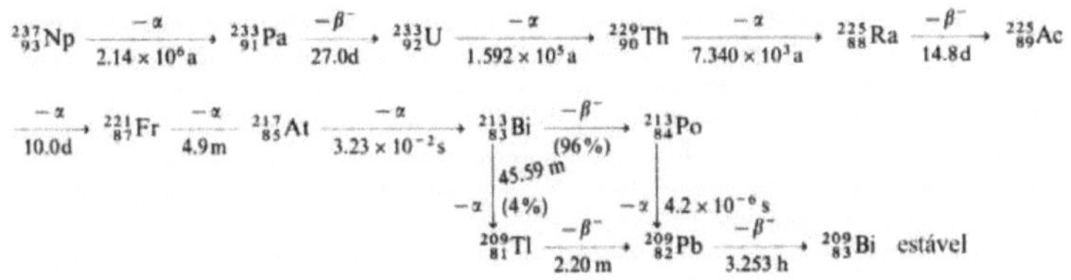

$$^{237}_{93}\text{Np} \xrightarrow[2.14 \times 10^6\,a]{-\alpha} {}^{233}_{91}\text{Pa} \xrightarrow[27.0d]{-\beta^-} {}^{233}_{92}\text{U} \xrightarrow[1.592 \times 10^5\,a]{-\alpha} {}^{229}_{90}\text{Th} \xrightarrow[7.340 \times 10^3\,a]{-\alpha} {}^{225}_{88}\text{Ra} \xrightarrow[14.8d]{-\beta^-} {}^{225}_{89}\text{Ac}$$

$$\xrightarrow[10.0d]{-\alpha} {}^{221}_{87}\text{Fr} \xrightarrow[4.9m]{-\alpha} {}^{217}_{85}\text{At} \xrightarrow[3.23 \times 10^{-2}s]{-\alpha} {}^{213}_{83}\text{Bi} \xrightarrow[(96\%)]{-\beta^-} {}^{213}_{84}\text{Po}$$

$$^{213}_{83}\text{Bi} \quad 45.59\,m$$
$$-\alpha \downarrow (4\%) \qquad\qquad -\alpha \downarrow 4.2 \times 10^{-6}\,s$$
$$^{209}_{81}\text{Tl} \xrightarrow[2.20\,m]{-\beta^-} {}^{209}_{82}\text{Pb} \xrightarrow[3.253\,h]{-\beta^-} {}^{209}_{83}\text{Bi} \quad \text{estável}$$

Figure 7 Schematic of the decay series from neptunium (not observed in nature). Abbreviations used: a = years; d = days; h = hours; m = minutes; s = seconds.

Table 3 The three natural decay series, better known as the actinium, uranium and thorium series. Abbreviations used: a = years; d = days; h = hours; m = minutes; s = seconds.

Série do actínio

Isótopo	Nome antigo	Meia-vida
$^{235}_{92}U$	**Actino-uran**	7.038×10^8 a
↓ α		
$^{231}_{90}Th$	Uran Y	25.5 h
↓ β		
$^{231}_{91}Pa$	Protactinium	3.276×10^4 a
↓ α		
$^{227}_{89}Ac$	Actinium[a]	21.77 a
↓ β		
$^{227}_{90}Th$	Radio-Actinium	18.72 d
↓ α		
$^{223}_{88}Ra$	Actinium X	11.43 d
↓ α		
$^{219}_{86}Rn$	Actinium-Emanation (Actinon)	3.96 s
↓ α		
$^{215}_{84}Po$	Actinium A[b]	1.78×10^{-3} s
↓ α		
$^{211}_{82}Pb$	Actinium B	36.1 m
↓ β		
$^{211}_{83}Bi$	Actinium C[c]	2.17 m
↓ α		
$^{207}_{81}Tl$	Actinium C"	4.77 m
↓ β		
$^{207}_{82}Pb$	**Actinium D (Chumbo actinio)**	∞

a $^{227}_{89}Ac \xrightarrow[1.2\%]{-\alpha} {}^{223}_{87}Fr \xrightarrow[21.8\ m]{-\beta} {}^{223}_{88}Ra$ (Actinium K)

b $^{215}_{84}Po \xrightarrow[5 \times 10^{-4}\%]{-\beta} {}^{215}_{85}At \xrightarrow[1.64^{-4}\ s]{-\alpha} {}^{211}_{83}Bi$ (Actinium B')

c $^{211}_{83}Bi \xrightarrow[0.32\%]{-\beta} {}^{211}_{84}Po \xrightarrow[0.52\ s]{-\alpha} {}^{207}_{82}Pb$ (Actinium C')

d $^{234}_{91}Pa \xrightarrow[0.15\%]{-\gamma} {}^{234}_{91}Pa \xrightarrow[6.75\ h]{-\beta} {}^{234}_{92}U$ (Uran Z)

Série do urânio

Isótopo	Nome antigo	Meia-vida
$^{238}_{92}U$	**Uran I**	4.468×10^9 a
↓ α		
$^{234}_{90}Th$	Uran X₁	24.10 d
↓ β		
$^{234}_{91}Pa$	Uran X₂[d]	1.17 m
↓ β		
$^{234}_{92}U$	Uran II	2.446×10^5 a
↓ α		
$^{230}_{90}Th$	Ionium	7.54×10^4 a
↓ α		
$^{226}_{88}Ra$	Radium	1600 a
↓ α		
$^{222}_{86}Rn$	Radium-Emanation (Radon)	3.825 d
↓ α		
$^{218}_{84}Po$	Radium A[e]	3.05 m
↓ α		
$^{214}_{82}Pb$	Radium B	26.8 m
↓ β		
$^{214}_{83}Bi$	Radium C[f]	19.9 m
↓ β		
$^{214}_{84}Po$	Radium C'	1.64×10^{-4} s
↓ α		
$^{210}_{82}Pb$	Radium D	22.3 a
↓ β		
$^{210}_{83}Bi$	Radium E	5.013 d
↓ β		
$^{210}_{84}Po$	Radium F	138.38 d
↓ α		
$^{206}_{82}Pb$	**Radium G (Chumbo rádio)**	∞

← existem alguns desvios →

Série do tório

Isótopo	Nome antigo	Meia-vida
$^{236}_{92}U$		2.342×10^7 a
↓ α		
$^{232}_{90}Th$	**Thorium**	1.405×10^{10} a
↓ α		
$^{228}_{88}Ra$	Mesothorium I	5.75 a
↓ β		
$^{228}_{89}Ac$	Mesothorium II	6.13 h
↓ β		
$^{228}_{90}Th$	Radiothorium	1.913 a
↓ α		
$^{224}_{88}Ra$	Thorium X	3.66 d
↓ α		
$^{220}_{86}Rn$	Thorium-Emanation (Thoron)	55.6 s
↓ α		
$^{216}_{84}Po$	Thorium A[g]	0.15 s
↓ α		
$^{212}_{82}Pb$	Thorium B	10.64 h
↓ β		
$^{212}_{83}Bi$	Thorium C[h]	60.60 m
↓ β		
$^{212}_{84}Po$	Thorium C'	3.0×10^{-7} s
↓ α		
$^{208}_{82}Pb$	**Thorium D (Chumbo tório)**	∞

e $^{218}_{84}Po \xrightarrow[0.02\%]{-\beta} {}^{218}_{85}At \xrightarrow[1.3\ s]{-\alpha} {}^{214}_{83}Bi$ (Radium B')

f $^{214}_{83}Bi \xrightarrow[0.04\%]{-\alpha} {}^{210}_{81}Tl \xrightarrow[1.32\ m]{-\beta} {}^{210}_{82}Pb$ (Radium C")

g $^{216}_{84}Po \xrightarrow[0.01\%]{-\beta} {}^{216}_{85}At \xrightarrow[\sim 3 \times 10^{-4}\ s]{-\alpha} {}^{212}_{83}Bi$ (Thorium B')

h $^{212}_{83}Bi \xrightarrow[36.2\%]{-\alpha} {}^{208}_{81}Tl \xrightarrow[3.1\ m]{-\beta} {}^{208}_{82}Pb$ (Thorium C")

Chemists are interested in the reactivity of the members of these series. Radium B" in Table 3 ($^{214}_{82}Pb$), for example, was analytically identical with lead and "radium C" ($^{214}_{83}Bi$) with bismuth. It is therefore possible to separate radium B from radium C using chemical methods - just apply the steps of the well-known "cation separation process". On the other hand, any attempt to separate radium B from lead, or radium C from bismuth, fails.

Remarkable in Table 3 are the three **"emanations"**, which comes from Latin and means to leak. Each series contains one of these, so they have been given the trivial names of Actinon, Radon and Thoron, respectively[8] . They all have a nuclear charge of 86 and are therefore isotopes of the noble gas radon ($^{219}_{86}Rn$, $^{222}_{86}Rn$ e $^{220}_{86}Rn$). They are responsible for the experimental fact that elements in their physical proximity become radioactive after a short period of time. We explain this "induced radioactivity" as follows: The emanation that comes out of the radioactive material spreads and settles on top of the inactive body, where

---

[8] The names vary little between languages, so we'll stick to those used in German or English.

it transforms under the emission of a particle α into polonium (a solid). This polonium, in turn, is unstable and radioactive.

To prove the emanations (= radon) and their decay product, the following experiment was carried out by pioneers *Ramsay* and *Soddy* in 1903: Radium salts emit radon, a gas which is introduced into a U-tube cooled by liquid N_2. The tube is then removed from the cold bath and coupled to a mass spectrometer. Only one signal is observed, that of radon. After a few days in the tube, however, two signals are noted, that of radon and that of helium.

End point of each of the decay series shown in Table 3 is inactive lead. According to the displacement laws, these products must have the atomic masses 206 ("uranic lead"), 207 ("actinic lead") and 208 ("thoric lead"), from the products of the uranium, actinium and thorium series, respectively. By analyzing the percentages of these isotopes in natural lead ore (average atomic mass of 207.2 u), we have an indicator for the relative occurrences of the three series throughout Earth's history.

Still analyzing the isotopes of Table 3 we realize that among the elements with order numbers >83 there are only three nuclides whose half-lives are high enough to sustain life on Earth (low estimates are 4.5 billion years): $^{232}_{90}Th$, $^{235}_{92}U$ e $^{238}_{92}U$. These are the elements responsible for the fact that bismuth (the last really stable nucleus), to this day, does not represent the end of the heavy elements in the Earth's crust. In fact, we found all the elements between $_{83}Bi$ and $_{90}Th$ and $_{91}Pa$. These elements exist, despite their shorter half-lives, because they are always reproduced from their parent substances. The elements shown in the three natural decay series are therefore in radioactive equilibrium ("secular equilibrium"; p. 59), where the relative share of each depends on the ratio between the speeds with which they were produced by the element above it, divided by the speed with which they decay into the elements just below it. Special thanks to the long half-life of $^{235}_{92}U$ with $t_{1/2}$ = 0.7 billion years we still find approx. 1/500 of the initial amount of $^{235}_{92}U$ in the ore, an almost inexhaustible reserve for humanity, to produce nuclear energy and $^{239}_{94}Pu$ fissile energy, among other applications (p. 118).

Outside the elements mentioned in Table 3 there are still 21 known natural elements with weak radioactivity. They are listed, in order of atomic mass, in Table 4. Note that they all have a long half-life (the type of decay and the half-life are shown in brackets).

Table 4 Radioactive isotopes that do not appear in the natural decay series of the heavy elements.

$^{10}_{4}Be$ (β⁻, 1.6· 10⁶ a)	$^{138}_{57}La$ (β⁻, 1.3· 10¹¹ a)	$^{176}_{71}Lu$ (β⁻, 3.3· 10¹⁰ a)
$^{40}_{19}K$ (β⁻, K, 1.28· 10⁹ a)[9]	$^{142}_{58}Ce$ (□, 5· 10¹⁶ a)	$^{174}_{72}Hf$ (□, 2.0· 10¹⁵ a)
$^{48}_{20}Ca$ (β⁻, >1.1· 10¹⁸ a)	$^{144}_{60}Nd$ (□, 2.1· 10¹⁵ a)	$^{187}_{75}Re$ (β⁻, 4.3· 10¹⁰ a)
$^{50}_{23}V$ (β⁻, >1.2· 10¹⁶ a)	$^{147}_{62}Sm$ (□, 1.06· 10¹¹ a)	$^{190}_{78}Pt$ (□, 6.1· 10¹¹ a)
$^{87}_{37}Rb$ (β⁻, 4.7· 10¹⁰ a)	$^{148}_{62}Sm$ (□, 7· 10¹⁵ a)	$^{192}_{78}Pt$ (□, ~10¹⁵ a)
$^{115}_{49}In$ (β⁻, 6· 10¹⁴ a)	$^{149}_{62}Sm$ (□, ~4· 10¹⁴ a)	$^{204}_{82}Pb$ (□, 1.4· 10¹⁹ a)
$^{123}_{52}Te$ (β⁻, 1.24· 10¹³ a)	$^{152}_{64}Gd$ (□, 1.1· 10¹⁴ a)	$^{209}_{83}Bi$ (□, 2.5· 10¹⁷ a)

Radioactivity and electromagnetic radiation

Both the He nuclei in the **α radiation** and the electrons in the **β radiation** are emitted from the decaying nuclei at high speed! The α particles come out at 5 to 10% of the speed of light, the e⁻ even come close to the speed of light! The new expressions in bold already suggest that we have to let go of the image that these fragments only have particle-like qualities. According to the particle-wave dualism, we can also give them wave qualities. They show, for example, the typical wave phenomenon known as interference. It is therefore correct to say that the decays α and β both have the characteristics of particles with restricted sizes (exact location, speed, impulse) and waves of infinite length (refraction, interference); which of these qualities we observe and take advantage of depends on the type of experiment carried out with the emitter α / □.

The radiation α and β is accompanied by a process that explains the high amount of energy released by the reactions in Equation 1 and Equation 2. This form of radiation does not involve matter like α and □, it is purely electromagnetic: **γ radiation**. It is a type of X-radiation, with a particularly short wavelength, in the order of $\lambda \approx 1$ pm (in comparison: the K radiation$_\alpha$ from molybdenum, widely used in diffraction tests on crystals, has a length of $\lambda = 0.71071$ Å $= 71.071$ pm. As the wavelength λ and the energy E *of* the radiation are inversely proportional, according to

---

[9] The symbol "K" stands for capture K, to be explained on p. 86. In this example, the process is slow and has taken place throughout the history of the earth. It can be blamed on the curious fact that argon, the element right before potassium in TPE, is heavier than the alkali metal potassium.

$$E = \frac{h \cdot c}{\lambda}$$

(with h = Planck's constant; c = speed of light in vacuum),

we can say that γ radiation, together with "cosmic radiation", are the most energetic forms of energy transport we know[10].

The effects of these three types of radiation, α, β and $\square$, are very different and led to the discovery of the phenomena of radioactivity. This can be seen by following the historical development of this branch of science.

In 1896, a few years after the discovery of X-radiation by *Wilhelm Röntgen,* the French researcher *Henri Bequerel* (1852 - 1908) observed that uranium salts caused photographic plates to darken when placed in their vicinity (p. 40). *Bequerel* concluded that an invisible and powerful radiation came from the uranium when he wrapped the photographic plate (containing silver ions in a gel base), occasionally leaving a key on top of the plate. After the plate was developed, the surprise: just where the key was, there was no darkening. The radiation from the sample then reached all parts, but did not pass through the metal of the key.

Two other phenomena about the radioactive sample:

- Nearby, the air becomes conductive;
- Certain salts, when placed near the uranium salt, begin to emit visible light, a phenomenon widely known as fluorescence (p. 40). These include: barium tetracyanoplatinate, $Ba[Pt(CN)_4]$, and zinc sulphide, ZnS.

Similar observations were soon made by the German physicist *Gerhard C. Schmidt* (1865 - 1949) studying thorium preparations.

But more spectacular discoveries were made in *Bequerel*'s laboratory. The main credit goes to the *Curie* couple (*Pierre Curie* 1859 - 1906 and *Marie Curie* 1867 - 1934), as their work led to the discovery of the strongly radioactive elements polonium and radium. This was an admirable achievement, given that the ore used, "Pechblende" or impure uraninite, contained only one Ra atom for every $3 \cdot 10^6$ U atoms (this is equivalent to 0.33 ppm, approximately). The Po content

---

[10] Of course, we can also apply the theory of particle-wave dualism to radiation $\square$. We would have the mass and velocity of a "particle". However, its mass, while stationary, would be zero. In general, we can say that the greater the mass, the more obvious and easier it is to take advantage of the phenomena as a particle. The lighter it is, the more obvious its wave-like qualities are.

was even lower than that of Ra, by a factor of 10^4. In 1902 *Marie Curie* reported that she had managed to isolate 0.1 g of $RaCl_2$ from two tons (!) of pechblende.

Soon afterwards, in 1899 - 1900, the English researcher *Ernest Rutherford* (1871 - 1937) and other physicists (*P. and M. Curie, H. Bequerel, P. Villard*) discovered that the radiation emitted by radioactive samples was not uniform. Their introduction into a magnetic field, shown in Figure 8led to the positively charged He nuclei being deflected in the opposite direction to the radiation β originating from the negatively charged electrons. The highly energetic electromagnetic radiation (= γ radiation) was not deflected at all. Persistence and refined tests finally led *Rutherford* and *Soddy* in 1903 to the bold hypothesis that the three types of radiation originate in spontaneous decay, in the instability of the nuclei of heavy elements.

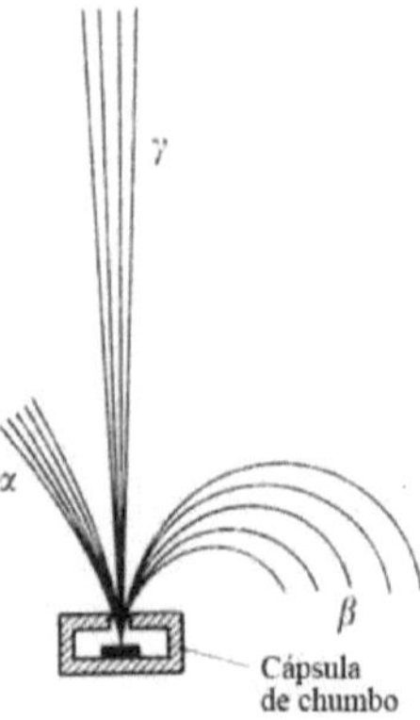

Figure 8 Deviation of radiation α, β and γ in the magnetic field (the field lines are perpendicular to the plane of the paper, the south pole above and the north pole below the plane of the paper).

Energy and radioactivity

All forms of radiation, □, β and □, emitted by radioactive samples, are very energetic. This is most obviously explained by the high speed (kinetic energy of the particles α and □) and high frequency of the electromagnetic radiation □. Let's look at some values:

- Particles α have an average of $6 \cdot 10^6$ eV (in the range between 4 and 9 MeV; read: mega electron volts). This energy corresponds to the energy of a proton (or an electron) after it has traveled through a potential field of 6 million volts!

- The energy of the particles β averages 1 MeV (in the range 0.02 to 4 MeV; most between 0.5 - 2 MeV).
- The γ rays are in the same region as the β radiation (mostly between 0.1 and 2 MeV).

Let's compare it with the energy phenomena most of us are familiar with:

- The thermal energy of the atoms in a white-hot body (this certainly implies temperatures above 2000 °C; see "black body" theory) is < 0.5 eV per atom.
- During the reaction of hydrogen with oxygen, a violent explosion, less than 3 eV are released for each molecule of water formed.
- Remember the series of electrochemical potentials E^0 , which indicate the power of oxidizing and reducing substances. These values, when multiplied by the number of electrons z transferred in the redox reaction, directly give the free energy $\Box G$ of the battery, as $\Box G = - z\,F\cdot\ \Box E$ ($F =$ Faraday's constant; 96,480 A s$^{..}$ mol^{-1} ; $\Box E =$ difference between the potentials of the reductant and oxidant in volts). The absolute values indicate that the energy released by redox reactions also does not exceed ~5 eV for each electron transferred (E values^0 from -2.5 V to +2.5V, approximately).
- The energy corresponding to visible light (400 nm $< \lambda <$ 750 nm) is 1.5 - 3 eV, with each photon $h\cdot\ \nu$ emitted.

Let's go back to radioactivity. For a particle α an initial velocity of $v =$ 17,000 Km s^{-1} is calculated using the relation:

$e \cdot U = \dfrac{m\cdot v^2}{2}$ ($e =$ elementary charge; $U =$ acceleration potential; $m =$ mass; v = velocity), which corresponds according to *Einstein*'s formula, $E = m\,c^2$, to about 6% of the speed of light.

The initial speed of the particle β is higher and reaches up to 99% of the speed of light. When you want to calculate this speed from the energy value, you can also use the relation $e \cdot U = \dfrac{m\cdot v^2}{2}$ but we have to include the "moving mass" of the electron. While its rest mass m is quite small ($m =$ 9.109534$\cdot$ 10^{-31} Kg), its moving mass m' is considerably larger, according to

$m' = \dfrac{m}{\sqrt{1 - v^2/c^2}}$ ($c =$ speed of light).

In the formula above, instead of 2 in the denominator, we insert the expression $\left[1 + \sqrt{1 - v^2/c^2}\right]$ It's easy to see that for zero velocity we'd go back to the original formula, *and* $U = (m\, v^2)/2$. However, the mass increases dramatically with the electron's velocity *v*, as the following values should show:

Electron energy	Speed *v*	Increased mass
[MeV]	[multiples of *c*]	$\Delta m = m' - m$
0,1	0,548	20%
0,5	0,863	98%
1	094	196%
5	0,996	1080%

This brings us to the estimated value of up to 99% of the speed of light.

Finally, the average energy of the rays γ оϕ ~1 MeV allows us to calculate their wavelength $\square$. The speed is naturally the speed of light, $v = c$, since this is electromagnetic radiation. So:

$$e \cdot U = \frac{h \cdot c}{\lambda} \; ; => \square \approx 10^{-10} \text{ cm} = 1 \text{ pm}.$$

To our surprise, a decaying nucleus <u>does not</u> emit uniform radiation α. For example, the radium $^{226}_{88}Ra$ in Equation 1 emits two types of radiation α whose energies are 4.78 and 4.59 MeV. The difference is $\square E = 0.19$ MeV - which corresponds precisely to the energy of the radiation γ (0.187 MeV) that accompanies the decay process $\square$. We can conclude that the difference in nuclear energy between radium and radon is expressed either in the form of just one very energetic α particle, or in the form of a slightly less energetic α, particle, plus a γ ray. But this also means that the nucleus, in any analogy to the electronic sphere, can only assume discrete energy levels. In the transition from one energy state to a lower state, energy is released, which in this case is radiation $\square$.

For clarity, let's represent the <u>high</u> energy state of the $^{222}_{86}Rn$ nucleus by a little star. Then we formulate

$$^{226}_{88}Ra \quad \rightarrow \quad ^{222}_{86}Rn \quad + \quad \alpha \quad + \quad 4{,}78 \text{ MeV} \quad \text{or}$$

$$^{226}_{88}Ra \quad \rightarrow \quad ^{222}_{86}Rn* \quad + \quad \alpha \quad + \quad 4,59 \text{ MeV}$$ then:

$$^{222}_{88}Rn* \quad \rightarrow \quad ^{222}_{86}Rn \quad + \quad 0,19 \text{ MeV}.$$

The 212 isotope of bismuth, $^{212}_{83}Bi$ or "thorium C", emits during its decay α to $^{208}_{81}Tl$ five types of radiation α, $\alpha\tau$ 6.084, 6.044, 5.762, 5.620 and 5.601 MeV. This leads to the conclusion that the thallium produced appears in its ground state or in one of the 4 excited states typical of this nucleus. Under radiation emission γ these latter nuclei can return to the fundamental nuclear state.

In this post, we want to remind you of the great importance of the energy states of the nucleus[11] , its magnetism[12] and the important analytical methods that take advantage of the phenomena within the nuclei[13] , allowing us to conclude the chemical structure of the compound examined - without going into these details, as there is already an extensive bibliographic base.

Υνλικε α radiation, β ραδιατιον has continuous "spectra". We expect the energy levels of the nucleus to be very close, or the emission from β το $\beta\epsilon$ accompanied by radiation from $\square$, also of continuous energy. It doesn't happen!

Radiation β is not accompanied by radiation $\square$, and we come into conflict with the principle of conservation of energy! A proposal to get around this dilemma came from *W. Pauli* (1931; theory perfected in 1934 by *E. Fermi*). He proposed that the nucleus that emits electrons still emits other fragments, almost free of charge and mass, whose kinetic energy compensates for the discrepancy

---

[11] *Gütlich P., Bill E., Trautwein A.* Mössbauer spectroscopy and transition metal chemistry: fundamentals and application. Springer, Berlin 2011, ISBN 978-3-540-88428-6.
Sharma V., Klingelhöfer G., Nishida T. Mössbauer Spectroscopy: Applications in Chemistry, Biology, and Nanotechnology. Wiley, 2013. ISBN 978-1-118-05724-7.
Rankin D., Mitzel N., Morrison C. Structural Methods in Molecular Inorganic Chemistry (Inorganic Chemistry: A Textbook Series). Wiley, 2013. ISBN 978-0-470-97278-6.
[12] Wikipedia: Nuclear magnetic resonance spectroscopy
Keeler J. "Understanding NMR Spectroscopy" (reprinted at University of Cambridge). University of California, Irvine. Available at https://www-keeler.ch.cam.ac.uk/lectures/Irvine/
Emsley J.; Feeney J.; Sutcliffe L. (1965). High Resolution Nuclear Magnetic Resonance Spectroscopy. Pergamon. ISBN 9781483184081.
Nuclear Magnetic Resonance Facility. Available at http://www.nmr.unsw.edu.au/usercorner/nmrhistory.htm
[13] *Roth K.* NMR-Tomography and -Spectroscopy in Medicine - An Introduction. Springer Berlin 1984. "NMR active nuclei for biological and biomedical applications". Open Medscience. Retrieved 2023-11-25.
Shah, N.; Sattar, A.; Benanti, M.; Hollander, S.; Cheuck, L. (January 2006). "Magnetic resonance spectroscopy as an imaging tool for cancer: a review of the literature". The Journal of the American Osteopathic Association. 106 (1): 23-27.
Siegel, R. W. (1980). "Positron Annihilation Spectroscopy". Annual Review of Materials Science. 10: 393-425.

between the energy observed instantaneously in the radiation β and its maximum energy. The new unknown species was called an **anti-neutrino**, its symbol $\bar{v}$. In the form of an equation:

$E_{radiationβ} + E_{anti-neutrino} = E_{max}$.

An example:

In the decay β of $^{14}_{6}C$ to $^{14}_{7}N$ there is a loss of mass in the nitrogen which can be transformed into energy using *Einstein*'s formula, $E = m\,c^2$. The following masses were observed:

14.3242 (C) - 14.003074 (N) = 0.000168 g/mol.

The equivalent in energy would be

0.000168· 931 = 0.155 MeV/atom.

The radiation particle β and the anti-neutrino share this kinetic energy in varying proportions.

$$E_{e^-} + E_{anti-neutrino} = 0{,}155 \;\; MeV$$

The more energy in the electron, the less energy in the anti-neutrino, and vice versa. Similarly, it is assumed that during the emission of positive electrons ("positrons"; and^+) there is simultaneous emission of neutrinos (symbol □). Direct proof of the anti-neutrino was provided in 1956, during the collision with protons, according to: $\bar{v} + p^+ \longrightarrow e^+ + n$.

The products here are a neutron and a positron. And shortly afterwards, in 1959, they also proved neutrinos, by reacting with neutrons according to $v + n \longrightarrow e^- + p^+$ producing a negative electron ("negatron"; e⁻) and a proton.

Interaction of radiation with matter

Due to their extremely high kinetic energies, α and β radiation can penetrate and even pass through matter. Even more powerful in this sense is γ radiation - which is not surprising when we remember the qualities of X-rays that pass through our bodies with ease; so imagine radiation 100 times more energetic...

The least penetrating are the particles α which, compared to the other forms of radioactivity, have a high mass and volume. On the other hand, a sheet of aluminum measuring just 1/200 mm can retain half of the radiation α emitted by radium. A 3 cm layer of air has the same effect. To stop the rays β from radio decay, you need a sheet of Al 100 times thicker, i.e. ½ mm thick. And the even more penetrating γ ραψσ need up to an 8 cm thick block of Al for the same effect!

On their way through matter, the rays □□□□□e γ then successively lose their energy, obviously because they interact with the atoms in the material (how - this will be explained below). It is therefore common to characterize this radiation, instead of indicating its energy in MeV, by its path through a defined medium where it loses half of its power. This range R *of* the radiation can be called, in analogy to the lifetime, the **"half-life thickness"** - always indicating the material penetrated and its physical state.

The range R *of* the rays α, in air of 0 °C and 1 atm can be calculated from their energy E, as follows:

$$E = 2{,}12 \cdot R^{2/3}.$$

This results in values from $R = 2.5$ cm (for $E_\alpha = 4$ MeV) to $R = 9$ cm (for $E_\alpha = 9$ MeV). For aluminum, these values are 0.02 and 0.06 mm, respectively. The radiation values β αρε then around 500 times greater: from 150 cm ($E_\beta = 0.5$ MeV) to $R = 850$ cm (for $E_\alpha = 2$ MeV) are the half-life thicknesses in air.

The final form of energy delivered to matter by any type of radiation is heat. Therefore, radium salts always have temperatures above ambient. One gram of radium generates approximately 400 J of heat in one hour.

As the kinetic energies of the particles released in radioactive decay are very high, the calorific value is equally gigantic. An example confirmed by the experiment: in the complete transition of 1 mole of radium into lead, 3400 million kJ of heat is generated - a figure that requires the complete combustion of 100,000 kg of coal; or the heat generated by the (violent!) reaction $H_2 + \frac{1}{2} O_2 \rightarrow H_2 O$ producing 250,000 kg of water.

These values can have important geological consequences. Calculations have revealed that solar radiation is not enough to keep the Earth at a constant temperature, the way we like it. To explain an average temperature of ~15 °C (see Table 7), we have to look for the causes, in the heat left over from its creation (>4.5 billion years ago) and - this is important in this context - the existence of radioactive ores that still exist. Only the current solar radiation, together with these two factors, can compensate for the heat that the globe loses to space. In the Earth's crust, radioactive elements are more abundant than in the Earth's mantle - even more so than in the Earth's core. But let's assume that the radioactive elements are evenly distributed, so a crust just 16 km thick would be enough to compensate for the Earth's natural heat loss, keeping its temperature stable.

There are also other geological phenomena that are certainly influenced by the heat generated by radioactive elements, such as geological cycles, the formation of mountains, the movement of continents and, of course, volcanism. All of these events take advantage of the immense heat generated in the depths by radioactive transformations, which has thus supported the melting of ore and movements throughout the existence of the globe.

There is also a technological application *for* this heat: SNAP (*System for Nuclear Auxiliary Power*) generators are maintenance-free energy sources. They are useful in aerospace, for example in satellites. They are small thermonuclear power stations in space (read more at: https://en.wikipedia.org/wiki/Nuclear_power_in_space).

When radioactivity penetrates matter, as we saw above, there are collisions with foreign atoms. These collisions almost exclusively involve the electronic layer - since the radius of the electronic sphere is much larger than the nucleus. The radius of the atomic nucleus is approximately 10,000 times smaller than the atomic radius, i.e. the same ratio as a grain of beans and an Olympic swimming pool. Therefore, when we talk about the radius of an atom, we always refer to the outer radius of the electronic layer (in other cases we talk about the ionic radius or the radius of the nucleus). The atomic radius[14] varies between 0.373 pm (for H) and 2.655 pm (for Cs). Remember that 1 pm = 0.01 Å = 10^{-12} m. In the general case, either electron activation or ionization can occur, while both events have important applications, as will be shown below. In rare cases, the nucleus is hit by radiation; this is where the events to be presented from p onwards occur. 69.

In the case of <u>activation,</u> the electrons are raised to a higher level. Remember that elevation corresponds to the absorption of defined energy levels, i.e. there are discrete energy levels where the electrons of an atom can be found (atomic emission and absorption spectra). For a short period of time, in the order of 10^{-8} s, the activated electron remains in the higher energy state and then returns to its original level. In this return to the normal state, energy is emitted by the atom which can take different forms. For example, this energy can be emitted in the form of visible light, hence we speak of "**fluorescence**". Another possibility is to trigger a chemical reaction in its vicinity (the breaking of a covalent bond,

---

[14] The atomic radius indicated here is defined as half the interatomic distance in the pure element at room temperature.

for example). These mechanisms explain the phenomenon that some radioactive substances are luminescent when in contact with air; others promote chemical reactions that would not occur in the absence of the radioactive source. Here are some examples and technological applications.

Radium salts often show a blue luminescence. This light comes from nitrogen in the atmosphere, in the sense that N_2 * $\rightarrow$ N_2 + h· $\square$. In the case of low radioactivity preparations, the phenomenon of luminescence requires the proximity of substances that are particularly susceptible to emitting light. One such substance is zinc sulphide, which emits a characteristic yellowish-green light. This phenomenon is still used today by mixing a small amount of a radioactive substance with ZnS powder. Historically, the first preparations were $^{228}_{90}Th$ Today, cheaper fission products from the nuclear reactor are used. These mixtures can be glued to the dial of a watch or on warning signs that become visible even in total darkness. Science also makes use of the easy luminescence of ZnS, as a screen with this salt can visualize particles $\square$. Each particle that hits the screen generates a small flash of light that allows it to be counted. The device is known as a scintillation counter or spintariscope. The scintiscope is used to determine the range of the radiation α, by determining the distance between the radioactive preparation and the screen where the flashes are still visible.

And a curiosity: using the spintariscope it is possible to determine, with good precision, the Avogadro number N_A . It measures the number of particles α emitted from a given amount of radium in a given time. One gram of radium emits 4.35· 10^{18} nuclei of He per year, which corresponds to 167 mm^3 of He gas. Estimating this volume of He-gas contained in 22.414 liters finally leads to the number sought:

$$N_A = \frac{4,35 \cdot 10^{18} \cdot 22,414}{167 \cdot 10^{-6}} = 6,08 \cdot 10^{23}.$$

The second effect of radioactive substances mentioned above is the <u>triggering of chemical reactions</u>. The most common reaction is the formation of ozone, because in the vicinity of any highly radioactive sample you can smell the characteristic odor of ozone (that penetrating smell you smell when you take an acrylic sweater over your hair, accompanied by little crackles). The same reaction occurs near silent discharges, for example in a transformer or *Siemens* ozonator. The source of this gas is oxygen in the atmosphere, according to

O_2 * + 2 O_2 $\rightarrow$ 2 O_3.

Another common phenomenon: water is cleaved into H_2 and O_2 by radioactivity. For example, an aqueous solution of radium chloride develops more than 30 mL of gases (H_2/O_2) per day and per gram of radium. In addition, hydrogen in the vicinity of radioactive samples is activated so strongly that it reacts voluntarily at room temperature with sulphur, arsenic and phosphorus to form H_2S, AsH_3 and PH_3 , respectively.

Let's also talk about the third possibility mentioned above, of <u>ionizing atoms or molecules </u>in the vicinity of radioactive substances. In this case, the outer electrons are not raised to higher levels, but are completely expelled from the atom. This ionization is a common phenomenon: a single 5 MeV particle α can generate up to 150,000 pairs of ions (cations and electrons) as it travels through the air! Similarly, β and γ rays also work. Radioactivity makes air conductive. This can be used to measure the radioactive intensity of samples. The radiation enters an ionization chamber, as shown in Figure 9. Inside the chamber, the air is being ionized. A current is established between two plates of a capacitor which can be measured. In most applications, only the radiation γ is measured by ionizing the air, by placing certain materials with a filter effect (blocking α and □) at the entrance to the chamber. The intensity of the radiation γ is indicated in terms of equivalents (in milligrams) of radium of the same intensity as the radioactive sample under analysis. This is the principle of the *Geiger* counter, which is widely used by executive authorities (nuclear disaster experts). But, in principle, the ionization chamber can be used to characterize all forms of radioactivity, α, β and □.

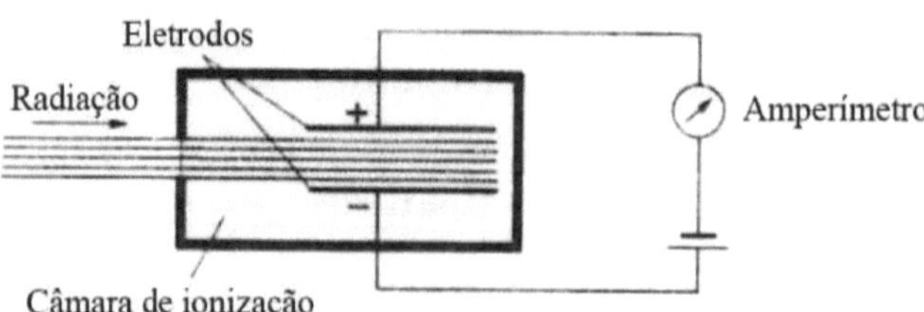

Figure 9 Ionization chamber, principle of the *Geiger* scintillation counter.

Instead of ionizing foreign atoms, electrons (= particles □) can be abstracted from the radioactive atom itself - especially by γ rays. The so-called **"electron conversion radiation"** is generated by this mechanism. Its energy naturally depends on the origin of the expelled electrons, whether from the K, L or M layer.

An example where ionization is coupled with a chemical reaction or a physical process, such as crystallization, is when rays hit colloidal silver bromide. An electron is abstracted from the bromide ($Br^- \rightarrow$ ½ $Br_2 + e^-$), which is then absorbed by the silver cation ($Ag^+ + e^- \rightarrow Ag\downarrow$). The uncharged silver, in turn, forms dark crystallites. These reactions are the basis of black and white photography. The radioactive rays then darken the photographic plate; this reaction, as already mentioned on p. 35has historical relevance, because *Bequerel*'s experiment led to the discovery of radioactivity.

There are also mineralogical phenomena that can be attributed to radioactivity. Rock salt, NaCl, can appear bluish - although it is quite pure. The blue color is due to free sodium atoms, formed under irradiation according to $NaCl \rightarrow Na +$ ½ Cl_2 . A similar explanation is given for the violet color of fluorite, CaF_2 , generated by Ca atoms via radiolysis. Or the grey color of "smoky quartz", SiO_2 , where Si atoms are believed to be the chromophore centers.

At the end of this chapter is the most pressing question we all have: what happens when radiation enters the human body? The answer is: all the phenomena described above can occur: electronic agitation, ionization and chemical reactions (either by providing the activation energy for new chemical bonds, or the breaking of existing ones). But medicine has found many uses for radiation, especially γ as it is the most penetrating of all. In principle, radiation also interacts with healthy tissue, but generally healthy tissue is more resistant than diseased tissue. With a sensible dosage, it is then possible to combat the damaged part (tumors; internal diseases and skin diseases). Although risky, radiation treatment is often the only way forward - especially in **cancer radiotherapy**.

In the past, medicine made use of natural γ radiators, such as $^{226}_{88}Ra$ or "meso-thorium I", $^{228}_{88}Ra$. These have been replaced by more accessible artificial isotopes, such as $^{60}_{27}Co$ e $^{137}_{55}Cs$ in radiotherapy, or $^{170}_{69}Tm$ for diagnostic purposes, to illuminate the human body.

Radioactive indicators (*Tracer*)

With radioactive isotopes we have a tool in hand to "mark" certain elements to be observed during their course in a chemical reaction or even in more complex organisms (our body, for example). This is due to the ease and high sensitivity of their detection. These isotopes emit radiation which, regardless of their chemical environment, can be detected; it accurately indicates minute traces of the element. This quality elevates radioactive isotopes, in the sense of indicators (English: *tracer*), above their non-radioactive relatives which can be used for the same purposes, but with less sensitivity.

A practical application in analytical chemistry is the characterization of poorly soluble heavy metal salts. In the example of lead chromate, $PbCrO_4$, it can be shown that a radioactive isotope makes it possible to determine the solubility product and even the surface quality of this finely distributed substance. It is not possible to determine the solubility of this salt by the standard method, i.e. by reducing a large volume of its saturated solution and weighing what has precipitated. The errors made in weighing are greater than the absolute quantity, due to the low solubility. The alternative is to mix a small amount of thorium B, $^{212}_{82}Pb$ in a still soluble starting salt, followed by precipitation with a soluble chromate. Lead chromate immediately precipitates. The part that remains in solution after filtering can be concentrated and its radioactivity compared with the radiation of the starting material. This allows the amount of soluble lead chromate to be calculated with high accuracy.

In a similar way, the surface area of precipitated lead sulphate can be determined by mixing thorium B, $^{212}_{82}Pb$ with $PbSO_4$. The relative activities of the radiations indicate the relationship $^{212}_{82}Pb_{superficie}/^{212}_{82}Pb_{solução}$. This ratio must be equal to the ratio of the atoms on the surface and in solution. Once the concentration of lead in solution has been determined, the number of lead atoms on the surface and thus the surface itself can be calculated. In addition to this method, the surface of solids and their changes during maturation, recrystallization or deformation can be determined when a mixed crystal is formed with a radioactive salt which, during its decay, releases "emanation" (expression explained on p. 34). 32). This gas can only be released when it is produced on the surface of the crystal. Its quantification then allows conclusions to be drawn about the surface; this is the emanation method according to *Hahn*.

Radioactive *traces* are important in solving biochemical problems. The important elements in biochemical processes are those at the beginning of the TPE (i.e. C, H, O, N, S, P), whose natural isotopes are not radioactive. In these cases, artificial isotopes must be used (p. 93) should be used as indicators.

The speed of decay

Half-life time

The rate of radioactive decay corresponds to a monomolecular reaction: the amount decayed in a given time, *dm/dt*, is directly proportional to the amount of radioactive material *m that* is available.

$$-\frac{dm}{dt} = k \cdot m \qquad\qquad\qquad \text{Equation 3}$$

According to Equation 3, the decay rate decreases over time and asymptotically approaches zero. The proportionality constant *k, in* this context, is called the **decay constant**. It indicates the amount of material that decays each second when starting from a single amount, i.e. for $m = 1$ we get $-\frac{dm}{dt} = k$.

For radium $^{226}_{88}Ra$ for example, the decay constant has the value $1.36 \cdot 10^{-11}$ s^{-1} ; from 1 g of radium $1.36 \cdot 10^{-11}$ g decays in one second. Due to the development of gigantic amounts of energy during radioactive processes, the value of k is literally independent of external conditions[15] .

The decay constant does not vary, no matter whether the radioactive material is examined at -273 °C or +3,000 °C; no matter whether the material is in the form of the element, as an ion or as part of a chemical compound.

The first observation that the activity of a radionuclide decreases by the same factor over equal periods of time - that is, it can be described by a fixed half-life - was published by *Rutherford* in 1900, observing the 220 isotope of the noble gas radon, $^{220}_{86}Rn$.

From the value of $k = 1.36 \cdot 10^{-11}$ s^{-1} for radium, it follows that during one year, from 1 g of Ra, the following decays occur $1{,}36 \cdot 10^{-11} \cdot 60 \cdot 60 \cdot 24 \cdot 365 = 0{,}000428$ *g* $(= 0.428$ mg$)^{16}$. This transformation mainly produces the long-lived isotope $^{210}_{82}Pb$ ("radium D") is formed. Each radium atom then loses 4 He atoms. The volume of helium produced in a year is therefore calculated to be $\frac{4 \cdot 22415 \cdot 0{,}000428}{226} = 0{,}17 \ cm^3/g$. This figure is in excellent agreement with the experiment.

---

[15] The only exception is the K capture, explained on p. 86.

[16] The simple calculation of the amount decayed, as done here, is admissible - although the loss in mass is not linear, but according to Equation 3 must be calculated via differential equation. However, from 1000 mg of starting material, only 4/10 mg is lost, i.e. the amount of Ra is practically constant.

After integrating Equation 3 between the mass limits m_0 (initial) and m_t (what's left after t seconds) this results in

$$ln\frac{m_0}{m_t} = k \cdot t \qquad\qquad \text{Equation 4}$$

The experimental determination of m_0 and m_t then leads directly to the decay constant k. The alternative would be to analyze the radiation intensity instead of the mass m, as these quantities are proportional. Once the constant k *has been* determined, Equation 4 can be used to calculate the time t required for mass loss $(m_0 - m)._t$

Equation 4, written in the form of $\frac{m_t}{m_0} = e^{-k \cdot t}$ that radioactive decay, represented by the loss of radionuclide mass, follows an exponential law. The time elapsed in which half of the original mass decays ($m_t = \frac{1}{2}\,m_0$) is called the **half-life time, $t_{1/2}$** . 50% of the atomic nuclei have been transformed into another nuclide - usually under the emission of ionizing radiation; this, in turn, may or may not be radioactive.

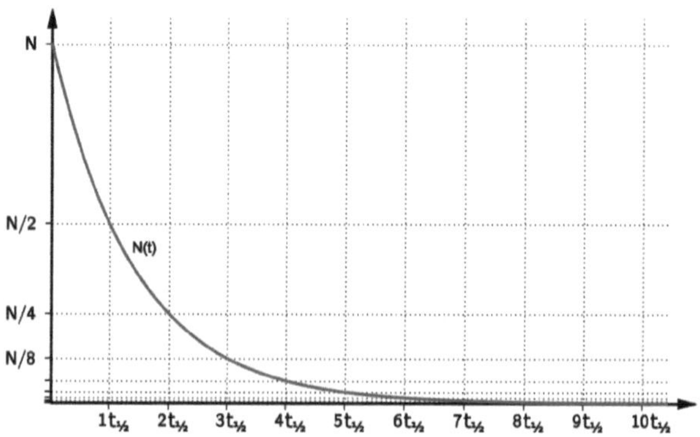

Figure 10 Exponential decrease of a quantity in relation to the initial value N. This could be, for example, the initial number of radioactive atomic nuclei in a sample. The velocity law makes it possible to calculate the concentration after time t: $N(t) = N \cdot e^{-t \cdot ln(2)/t_{1/2}}$.
Source: https://de.wikipedia.org/wiki/Halbwertszeit

Equation 4 results:

$$t_{1/2} = \frac{0{,}693}{k} \qquad\qquad \text{Equation 5}$$

Most texts refer to the half-life time $t_{1/2}$, rather than the decay constant k, as a period is more figurative. For $^{226}_{88}Ra$ this period, using Equation 5, is

$$t_{1/2} = \frac{0,693}{1,36 \cdot 10^{-11}} = 5,11 \cdot 10^{10} s$$ which corresponds - as a year has $60 \cdot 60 \cdot 24 \cdot 365 = 3,15 \cdot 10^7 s$ - to 1622 years. Whatever amount of radium it is, after exactly 1622 years it decays by half. In other words, the decrease in radiation from 100% to 50%, from 50 to 25, from 25 to 12.5%, takes the same amount of time as 1622 years.

One more important detail: the chemical environment of the radioactive element only influences the electronic sphere, but not the nucleus. The half-life thus <u>does not</u> depend on the form the element takes: it can be a piece of metal, a solid salt, a rock, a solute ion in water - it doesn<u>'t</u> matter. The element will always have the same $t_{1/2}$.

And even more impressive: the element can be stored at any temperature - its half-life remains constant. This becomes clear when comparing the gigantic energies involved in a nuclear reaction (in the order of MeV), with the thermal energies (in the order of a few eV; compare p. 37).

As seen in Table 3 natural isotopes have the most diverse half-lives; $t_{1/2}$ ranges from 1/10,000th of a second (for thorium C'; $^{212}_{84}Po$) and up to 10 quadrillion years ($\approx 10^{26}$ s; in the case of $^{204}_{82}Pb$). An overview of the natural nuclides and their half-lives is given in the nuclide map in Figure 11.

The half-lives of all radionuclides can be found in the isotope list[17] , where they are referred to along with other data on nuclide maps. A widely used printed collection is the "Karlsruhe Nuclide Map"[18] . A nuclide map from the Korean Atomic Energy Research Institute is available online[19] .

---

[17] https://en.wikipedia.org/wiki/Table_of_nuclides
[18] *J. Magill, G. Pfennig, R. Dreher, Z. Sóti.* Karlsruher Nuklidkarte. 8th ed. 2012. Nucleonica GmbH.
[19] https://atom.kaeri.re.kr/nuchart/

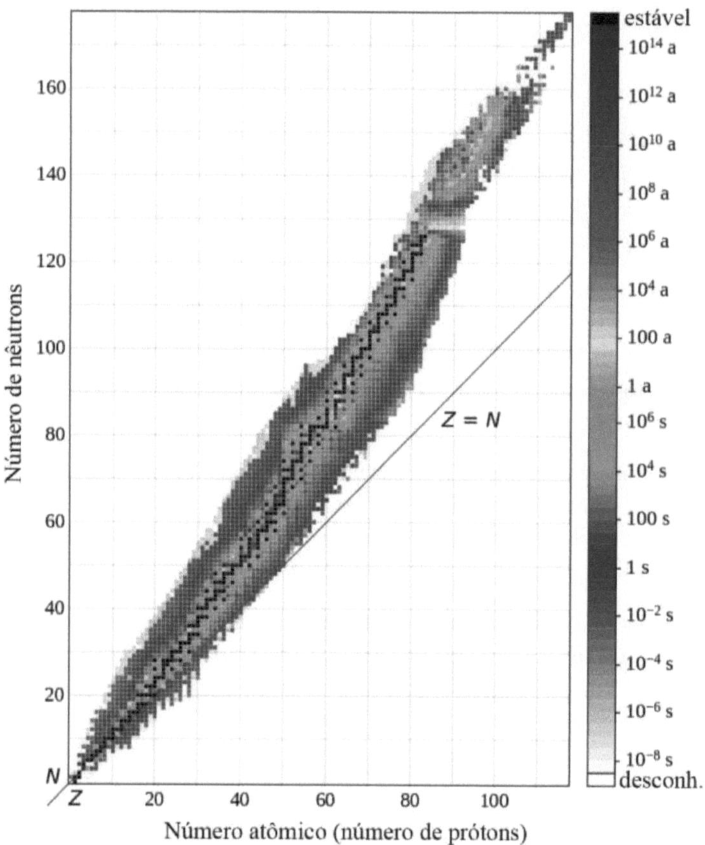

Figure 11 Map of nuclides with half-life times, represented by the colors according to the scale on the right. Source: https://de.wikipedia.org/wiki/Halbwertszeit

On the abscissa of Figure 11 that the last black dot (representing the stable nuclei) is the element $Z = 83$, so bismuth. The entire black string on this graph is called the **stability line**. It was only at the end of the 20th century that some nuclides that had previously been considered stable were "unmasked" as extremely long-lived radionuclides, for example, [149] Sm, [152] Gd (both lanthanides), Hf, [174180] W and [209] Bi, with half-lives of a few trillion years. With such long half-lives, radioactivity is correspondingly low and can only be detected with great effort.

50

Pure and mixed elements

There are more mixed elements than pure ones. A pure element, also known as an anisotope element, is a chemical element of which there is only one stable or extremely long-lived isotope on Earth (before human intervention). In the natural occurrence of a pure element, all its atoms have the same atomic mass and also agree in all other properties. All other elements, where there are two or more stable isotopes, are called mixed elements.

There are 22 pure elements, of which three (bismuth, thorium and plutonium) are unstable:

Beryllium, Fluorine, Sodium, Aluminum, Phosphorus, Scandium, Manganese, Cobalt, Arsenic, Yttrium, Niobium, Rhodium, Iodine, Cesium, Praseodymium, Terbium, Holmium, Thulium, Gold, Bismuth, Thorium and Plutonium.

Although bismuth, thorium and plutonium are not stable, there are still natural occurrences due to their very long half-lives, so they should also be counted with the pure elements. In the case of plutonium, however, the amount of the most stable isotope^{244} Pu that has remained since the formation of the Earth is so small that the amount of plutonium now produced by humans (mainly producing the isotopes^{238} Pu to^{242} Pu) will probably exceed the amount of primordial plutonium on the planet by orders of magnitude, so that^{244} Pu is no longer the most abundant plutonium isotope on the planet.

All the pure elements, except beryllium, thorium and plutonium, have odd atomic numbers. The fact that the mass numbers of most pure elements are odd led to the "*Mattauch* isobaric rule" (1934), an empirical rule of radiochemistry. It states that stable isobars (= nuclides with the same mass number) from neighboring elements do not occur in TPE; the difference in the atomic numbers of two stable isobars is always greater than one.

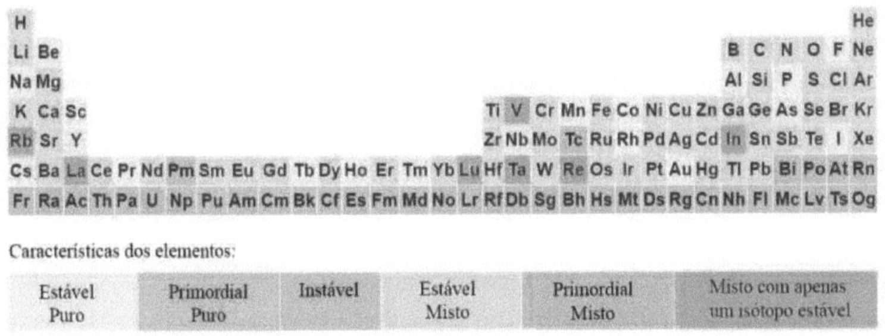

Figure 12   Pure, mixed   and   primordial   elements.
Translation of https://de.wikipedia.org/wiki/Reinelement

In nature, other isotopes of pure elements are also formed in small quantities due to nuclear transformation by cosmic radiation and spontaneous fission. Some of the unstable elements in Figure 12 highlighted in gray, also appear in dashes.

<u>Meaning of pure elements</u>. In order to technically separate a mixed element into its isotopes, it is often necessary to treat it in its elemental (unbound) form, since many other chemically bound elements also have isotopes and contribute to differences in the mass of molecules. Chemical compounds with pure elements do not have this disadvantage. For example, uranium isotopes can be separated in the chemical form of uranium hexafluoride (p. 122). This is only possible because fluorine is a pure element, so the mass differences in the molecules UF^{235}_6 and UF^{238}_6 are due solely to the uranium isotopes.

The other extreme are elements of which there are a large number of different stable isotopes. These are precisely the elements whose nuclei have higher than expected stability ("magic number of protons": 2, 8, 20, 28, 50, 82, 126), which stand out because they exist as a mixture of many natural isotopes. Examples are calcium ($Z = 20$) with 6 stable natural isotopes; nickel ($Z = 28$) with 5 isotopes; the elements Ba, Mo, Ru, Nd, Sm, Yb, Os and Hg with 7 isotopes each; the champions are xenon with 9 and tin with 10 stable isotopes.

It is important to know that the relative quantities of these isotopes do not change, regardless of where the element was found (at the bottom of the sea, in high mountains, in meteorites or on the moon).

The magic numbers on the neutrons, by the way, are the same: 2, 8, 20, 28, 50, 82, 126. Evidently, in the nucleus, in analogy to the electronic sphere, there is a principle of order in layers: when they are full, they are especially stable.

Although they are generally complicated physical systems, atomic nuclei are often represented simply as spherical agglomerations of protons and neutrons. But many atomic nuclei are not round, but have significant deformations. This is due to the interaction of forces acting on these small, massive centers that accumulate almost all matter: the strong nuclear force holds the protons and neutrons together, while the electromagnetic force ensures that positively charged protons are repelled. The general rule applies: the less stable an atomic nucleus is, the more deformed it is. Nuclei can then be flattened like a disk or elongated like a rugby ball. An international research group from the Canadian research center *Triumf* discovered in 2019 that even particularly stable nuclei are sometimes not spherical.

However, we must not forget that the half-life time is only a statistical average. The transformation time of an individual atomic nucleus cannot be predicted; only the probability of transformation in a time interval can be predicted.

The probability of a single core under consideration becoming

- in the first half-life is 50%,
- in two half-lives is 50% + 25% = 75%,
- in three half-lives is 50% + 25% + 12.5% = 87.5% etc.

The time $t_{1/1000}$, after which only one millionth of the original amount of the radioactive element is left, i.e. where 99.9% has already decayed, can be calculated with Equation 4:

$$t_{1/1000} = \frac{6,908}{k} = 9,9 \cdot t_{1/2}$$

This period is then approximately 10 times the half-life, from Equation 4. Returning to our example: after $1622 \cdot 10 = 16,220$ years a sample of $^{226}_{88}Ra$ has practically disappeared.

By analogy, we calculate the periods in which 90% are decayed ($t_{1/10} = \frac{2,303}{k} = 3,3 \cdot t_{1/2}$) and in which 99% are decayed ($t_{1/100} = \frac{4,606}{k} = 6,6 \cdot t_{1/2}$).

In general: after n half-lives, the amount of $\left(\frac{1}{2}\right)^n$ of the radioactive isotope remains.

n	1	2	3	4	5	6	7	8	9	10
$\left(\dfrac{1}{2}\right)^n$	$\dfrac{1}{2}$	$\dfrac{1}{4}$	$\dfrac{1}{8}$	$\dfrac{1}{16}$	$\dfrac{1}{32}$	$\dfrac{1}{64}$	$\dfrac{1}{128}$	$\dfrac{1}{256}$	$\dfrac{1}{512}$	$\dfrac{1}{1024}$

Of the elements with half-lives of 1/8 or 1/9 or 1/10 of the Earth's maximum age (stipulated as $< 6 \cdot 10^9$ years) we can say that their original quantities m_0 were 250 or 500 or 1000 times greater than they are today.

- Of an element with a half-life of $t_{1/2} = 3 \cdot 10^8$ s ($n = 20$) today there are only 10^{-6} parts of the original quantity left.
- Of an element with a half-life of $t_{1/2} = 2 \cdot 10^8$ s ($n = 30$) today there are only 10^{-9} parts of the original quantity left.
- Of an element with a half-life of $t_{1/2} = 1.5 \cdot 10^8$ s ($n = 40$) today there are only 10^{-12} parts of the original quantity left.

The transuranic elements, whose half-lives are in this order of magnitude, are now almost completely extinct. Of course, this doesn't apply to those elements that are always reproduced from their long-lived parent elements (see Figure 11)!

On the other hand, elements with half-lives 10 times the age of the Earth still have more than 93% of their original quantity left today:

For $n = 1/10$ results $\left(\frac{1}{2}\right)^n = \frac{1}{1,07} \approx 93\%$.

Closely linked to the half-life of a radionuclide is its **specific activity**, i.e. the activity per mass, expressed in **becquerels per milligram, Bq/mg**. One becquerel (Bq) indicates the number of atoms that decay per second. The relationship between specific activity and half-life is inversely proportional: the shorter the half-life, the higher the activity for a given amount of substance, and vice versa.

A Table 5 contains some examples. Only the mass of the radionuclide itself is considered in the numerical values. In practice, it is more likely that specific activities are related to the respective mixture of natural isotopes or to the total sample material.

Table 5 Relationship between half-life and specific activity of some technologically/ecologically relevant nuclides.

Isotope	Half-life time	Specific activity
$^{131}_{53}I$	8 d	4,600,000,000,000 Bq/mg
$^{3}_{1}H$	12,33 a	370,000,000,000 Bq/mg
$^{137}_{55}Cs$	30 a	3,300,000,000 Bq/mg
$^{239}_{94}Pu$	24.110 a	2,307,900 Bq/mg
$^{235}_{92}U$	703.800.000 a	80 Bq/mg
$^{238}_{92}U$	4.468.000.000 a	12 Bq/mg
$^{232}_{90}Th$	14.050.000.000 a	4 Bq/mg

Measuring radioactive half-lives

Due to the very different absolute values of the isotopes, different methods are needed to measure the half-life times.

- In an average range of half-life times varying from seconds to days, it is possible to directly follow the decrease until half of the activity is reached.

- Very long half-lives are measured by counting the decays in a time interval of a known quantity of the substance; instead of the time $t_{1/2}$ the decay constant k *is* then observed. The exact amount of the radionuclide can be determined, for example, using a mass spectrometer. Using this method, the half-life of the iron isotope $^{60}_{26}Fe$ of $t_{1/2} = 2.5 \cdot 10^6$ years was measured with an accuracy of 2%. [20]

- For very short half-lives there are techniques that determine where decay occurs, when the radioactive atom (or molecule) passes through a series of detectors at a known speed[21] . It is also possible to use the "*Geiger* and *Nuttall* rule" (1911). Empirically, they found that the decay constant k and the range R *of* particles α in air are logarithmically related. In the case of the uranium series (Table 3) the equation applies:

$$log\ k = -37{,}7 + 53{,}9 \cdot log\ R.$$

Hence, the decay constant of the $^{214}_{84}Po$ (radio C'), whose range is $R = 6.60$ cm, should be around 10^6 s^{-1} .

- Extremely short half-lives, for example of excited nuclear states in the range of 10^{-22} to 10^{-16} seconds, can be measured by means of the **decay width** Γ of the resulting radiation. "Width" refers to the half-width of the relevant maximum (peak) in the graph of the excitation function (cross-section of the effect as a function of the energy of the center of mass). The width therefore directly shows the energy uncertainty of the state. The smaller the width (decay) of a peak, the longer the lifetime of the particle state, and vice versa. According to quantum mechanics, an unstable state cannot have a well-defined energy; only states that do not change have fixed energies. The decay width has the dimension of an energy, it is usually indicated in electronvolts.

Radioactive balance

Radium $^{226}_{88}Ra$ emits radiation α and turns into radon $^{222}_{86}Rn$. From M_{Ra} moles of radium are produced every second

[20] *A. Wallner* et al. Physical Review Letters **114** (2015), 041101
[21] *E. B. Paul.* Nuclear and Particle Physics. North-Holland 1969, p. 47-49.

$$k_{Ra} \cdot M_{Ra} = m_{Rn} \text{ moles of radon.} \qquad\qquad \text{Equation 6}$$

But the radon formed is not a stable isotope. From M_{Rn} moles of this gas that are available in a given time, decay

$$k_{Rn} \cdot M_{Rn} = m'_{Rn} \text{ moles,} \qquad\qquad \text{Equation 7}$$

under radiation emission α and the formation of $^{218}_{84}Po$ ("radium A"). At the start of the test, the amount M_{Rn} of radon will increase, as $m_{Rn} >$ m'$_{Rn}$; more radon is formed than disappears. Gradually, however, the amount of radon M_{Rn} and thus also the amount that decays increases, until an equilibrium state is reached, where

$$m_{Rn} = m'_{Rn} \qquad\qquad \text{Equation 8}$$

From this moment on, the amount of radon no longer changes, as the number of atoms that form is identical to the number of atoms that decay. This situation is known as **radioactive equilibrium**. Inserting equations 6 and 7 into Equation 8 leads to the equilibrium condition,

$$k_{Ra} \cdot M_{Ra} = k_{Rn} \cdot M_{Rn}.$$

Also respecting Equation 5:

$$\frac{M_{Ra}}{M_{Rn}} = \frac{k_{Rn}}{k_{Ra}} = \frac{t_{1/2,Ra}}{t_{1/2,Rn}} \qquad\qquad \text{Equation 9}$$

In other words: in radioactive equilibrium, the quantities of isotopes are in the same ratio as their half-life times and inversely proportional to their decay constants.

In our example we have inserted $2.10 \cdot 10^{-6}$ s^{-1} for k_{Rn} and $1.36 \cdot 10^{-11}$ s^{-1} for k_{Ra} . Radium and radon are in equilibrium when there are 155,000 times more Ra atoms than Rn.

Before reaching equilibrium, the activity of a certain isotope increases or decreases - depending on whether the half-life of the daughter element is shorter or longer than that of the parent element. This is why *Rutherford* and *Soddy* observed in 1902 that a freshly prepared thorium sample increased in quantity and activity over time.

A little more history and how the unit of radioactive activity, **Curie (Ci)** changed during the 20th century. From 1910 onwards, the amount of radon that is in equilibrium with 1 g of radium, i.e. approximately 0.6 mm^3 , was called "1 Curie". This definition was later changed to mean that the activity of a 1 Curie isotope corresponds to the activity emitted by 1 g of radium in a unit of time. However, over the years the sensitivity and accuracy of the analytics (decay constant and molar mass of radium) increased and the value of 1 Ci based on radium activity had to be corrected several times. Finally, in 1950 it was defined that 1 Ci is that quantity of an isotope that emits exactly $3.7 \cdot 10^{10}$ particles per second.

Acronym	Extension	Particles emitted per second
1 mCi	milli-curie	$3.7 \cdot 10^7$
1 $\square$Ci	micro-curie	$3.7 \cdot 10^4$
1 MCi	Mega-curie	$3.7 \cdot 10^{16}$

According to Equation 8, when a system is in radioactive equilibrium, the activities of the mother and daughter isotopes, measured in Curie, are identical. Because remember: each daughter atom that is born causes the mother to die; and the birth and death rates of the daughters are the same, so the activities (= deaths) of mother and daughter are the same.

Equation 9 is valid not only for neighboring isotopes, but can be extended to all isotopes in a decay series. Regardless of the relative position of the members - they can be neighboring or remote isotopes - radioactive equilibrium can be established. This can be seen, for example, in the atomic ratio [radium : uranium] in the mineral pechblende, which is constant ($M_{Ra} : M_U = 3.60 \cdot 10^{-7}$); since uranium and radium have no mother-daughter relationship, but are very distant in the series.

Once you know the ratio [M_A : M_B] of two isotopes in the radioactive equilibrium and one of the decay constants (or half-life time), the constant of the other isotope can be calculated using Equation 9. This is the way to calculate especially long half-lives, for example. The decay constant of long-lived elements is difficult to observe directly because their activity is lower than the accuracy of the instrument. For example, from the ratio [$M_{Ra} : M_U$] = $3.60 \cdot 10^-$

[7] and the activity of radium, which is easy to measure, the half-life of uranium was calculated:

$$\frac{1622}{3,60 \cdot 10^{-7}} = 4,51 \cdot 10^9 \text{ anos.}$$

Another example of using Equation 9:

Based on the knowledge of the half-life times of $^{238}_{92}U$ ($t_{1/2} = 4.51 \cdot 10^9$ years) and of $^{230}_{90}Th$ ($t_{1/2} = 7.52 \cdot 10^4$ years), a sample in the state of radioactive equilibrium contains, for each mole of uranium,

$$\frac{7,52 \cdot 10^4}{4,51 \cdot 10^9} = 1,67 \cdot 10^{-5} \text{ mol of thorium, or by mass: for every gram of uranium}$$
it has

$$1,67 \cdot 10^{-5} \cdot \frac{230}{238} = 1,61 \cdot 10^{-5} \text{ grams of thorium. In other words: 16 mg of}$$
thorium next to 1 kg of uranium.

Determining the age of minerals and rocks

Among the practical applications of radioactive equilibrium and decay rate according to Equation 4, we must not forget the determination of age in geological periods. This method is mainly applied to minerals, but later on we will also talk about the analysis of organic material via radioactive carbon, which, however, indicates the age of a few thousand years, whereas here we are talking about a few billion years.

A radionuclide is called "**primordial**" (Latin: of the first order) if it was already present when the Earth was formed and has not yet completely decayed. It therefore occurs in nature without being supplied by natural or technical processes.

Assuming that the Earth is 4.6 billion years old, the half-life of a nuclide must be more than 50 million years for there to be any possibility of detection. This means that a maximum of 288 nuclides are possible. According to current knowledge, these are divided into 253 stable and 35 primordial. The list, ordered by decreasing half-lives, ends with

- Platinum-190 ($t_{1/2} = 650$ billion years, 99.5%),
- Samarium-147 ($t_{1/2} = 106$ billion years, 97%),
- Lanthanum-138 ($t_{1/2} = 105$ billion years, 97%),
- Rubidium-87 ($t_{1/2} = 49$ billion years, 94%),

- Rhenium-187 ($t_{1/2}$ = 41 billion years, 93%),
- Lutetium-176 ($t_{1/2}$ = 38 billion years, 92%),
- Thorium-232 ($t_{1/2}$ = 14 billion years, 80%),
- Uranium-238 ($t_{1/2}$ = 4.47 billion years, 49%),
- Potassium-40 ($t_{1/2}$ = 1.25 billion years, 7.8%),
- Uranium-235 ($t_{1/2}$ = 704 million years, 1.08%),
- Plutonium-244 ($t_{1/2}$ = 80 million years, $5\text{-}10^{-16}$ %).

The percentage figures indicate the proportion that is still present after 4.6 billion years from the original 100%. The plutonium isotope ^{244}Pu (half-life of 80 million years) was detected as a radionuclide in 1971 using mass spectrometry. Its half-life has expired more than 57 times during the age of the Earth, making it the most transient primordial nuclide. Its original concentration was approximately $1.5 \cdot 10^{17}$ times higher than today. Its mass fraction in some ores is 10^{-18} .

Primordial radionuclides are usually mixed with other, sometimes stable, isotopes of the same element. Other important primordial nuclides besides those already mentioned above include: ^{190}Pt, Pb,^{204209} Bi and^{40} K. The latter - contained in all living organisms - has a half-life of 1.28 billion years.

Distinguishing between stable and primordial (radio) nuclides is difficult due to their long half-lives. For some primordial nuclides - in particular^{235} U,^{238} U and^{232} Th - the decay product ("daughter nuclide") is not stable, but also radioactive. With the nuclides mentioned above, this is the case over several generations of daughter nuclides. If, as with the aforementioned nuclides, the daughter nuclides have shorter half-lives than the parent nuclide, then after a long time an "equilibrium secular " is established in which the activity of the daughter nuclides is equal to the activity of the parent nuclides. Undisturbed rocks containing uranium or thorium therefore always contain all the daughter nuclides of the uranium-radium and uranium-actinium decay series or of the thorium decay series. This makes uranium-rich ores more radioactive than pure uranium. *Marie Curie* and her husband *Pierre* realized this fact when they compared solutions of uranium salt (already available at the time), with pechblende from a region in central Germany. This led them to correctly conclude that pechblende must also contain other radioactive elements. The *Curie* couple managed to detect one of these elements, radium, for the first time, which won them the Nobel Prize in 1911.

On some nuclide maps, the primordial radionuclides are specially marked, e.g. B. on the Karlsruhe nuclide map [18] there is a black bar at the top of its color field.

The longest-lived primordial nuclides are those that can only transform through the rare process of double beta decay, while single beta decay is not possible for them. The "record holder" is the aforementioned[128] Te with a half-life of 7.2-10^{24} years (see list of isotopes/atomic number 51 to atomic number 60); that's about 520 trillion times the age of the universe.

^{232} Th, U,^{238235} U and^{40} K are of practical importance - in technical terms or as part of natural exposure to terrestrial radiation. The radioactivity of longer-lived nuclides is meaningless outside of academic considerations and extremely sensitive measurements. Shorter-lived primordial nuclides are no longer found in relevant quantities on Earth.

As can be seen in the natural decay series of uranium (Table 3), the uranium atom has a tendency to decay producing the inactive $^{206}_{82}Pb$ ($k = 0$; $t_{1/2} = \infty$) which is at the end of the series. But this happens very slowly.

The U $\rightarrow$ Pb transition in geological times can be used to determine the age of uranium ore. Before showing how this is done, however, we should note that the average atomic mass of lead is 207.2 u. This indicates that lead is not a pure element, but a mixture of different isotopes. Of these, only $^{206}_{82}Pb$ is of interest to us, as it is the final element in the uranium series.

If, by chance, the natural mixture of lead (which has not been produced by radioactive decay) is present, the proportion of lead of non-radioactive origin can be discovered by the presence of $^{204}_{82}Pb$. As the composition of the mixture $[^{204}_{82}Pb : ^{206}_{82}Pb : ^{207}_{82}Pb : ^{208}_{82}Pb]$ in natural lead is known, this part must be subtracted from the total lead in the sample, in order to only calculate with the isotope $^{206}_{82}Pb$ that originated from uranium. We will then calculate the number of years it took to produce this lead from uranium. This is shown in the example of the Monogoro ore (Africa), where a ratio $[^{206}_{82}Pb : ^{238}_{92}U] = 0,107$. In other words, for every mol of uranium, 0.107 mol of lead is found. This^{206} Pb was formed at the cost of^{238} U. To establish the original U content, this part must be added:

$$\frac{m_{U,0}}{m_{U,t}} = \frac{1+0,107}{1} = 1,107$$

As $^{238}_{92}U$ has a decay constant of $k = 1.54 \cdot 10^{-10}$ per year, Equation 4 reveals its age:

$$t = \frac{ln\, 1,107}{1,54 \cdot 10^{-10}} = 660 \cdot 10^6 \text{ anos}$$

This result must be systematically small, because we don't take into account the fact that all the other isotopes in the series are present in the sample. A correction factor must also be introduced due to thorium, which naturally always accompanies natural uranium and also decays.

Similarly, the following atomic ratios can be analyzed to reveal the age of ores:

$[^{207}_{82}Pb : ^{235}_{92}U]$;

$[^{208}_{82}Pb : ^{232}_{90}Th]$; so

$[^{206}_{82}Pb : ^{207}_{82}Pb : ^{208}_{82}Pb]$.

Of the minerals that have been analyzed by this method, we know that the geological ages vary between 60 million years old, a uranite from the Upper Cretaceous formation, and 3500 million years old, discovered in a Precambrian uranite. It was concluded that the Upper Cretaceous formation is 60 million years old and the Precambrian is 3.5 billion years old. In comparison: the entire age of the Earth can be estimated at 4.6 to 6 billion years, and that of the universe at 20 billion years, roughly.

Although this theory is daring, we can go one step further to draw conclusions from the radioactive balance. We then assume that God created uranium, with the isotopes $^{235}_{92}U$ e $^{238}_{92}U$ isotopes, in equal parts. We further assume that this happened at the beginning of the Earth's existence. As we know that the half-life times are $6.96 \cdot 10^8$ and $4.51 \cdot 10^9$ years, respectively, then we can calculate the time x where there was equality between the isotopes:

$$99,2739 \cdot 2 \cdot e^{x \cdot 4,51 \cdot 10^9} = 0,7205 \cdot 2 \cdot e^{x \cdot 6,96 \cdot 10^8} ; => x = 5.85 \cdot 10^9 \text{ years.}$$

From this calculation, the Earth must be around 6 billion years old. A Figure 13 shows this calculation graphically, where time t is placed as the abscissa on a linear scale and the quantities of the isotopes in grams are represented on the ordinate, on a logarithmic scale.

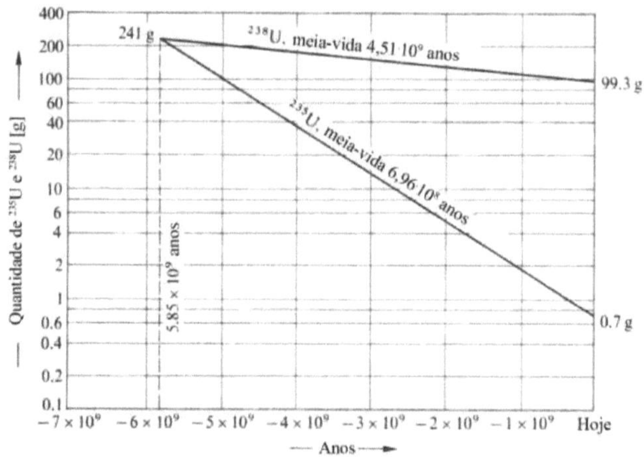

Figure 13 Graphical determination of the age of the Earth.

Incidentally, it's interesting that meteorites that have fallen to Earth contain uranium isotopes in the same ratio as the Earth itself. They must therefore be the same age and have the same origin...

There is also an alternative to the lead method for determining the age of uranium ores and meteorites - although this is less reliable. Instead of analyzing the amounts of lead, it offers the quantification of the helium gas that was released during the decays and which was trapped inside the rock. One gram of uranium, together with its decay products, produces $1.1 \cdot 10^{-7}$ cm^3 of He in one year. As we can see in Table 3an atom of $^{235}_{92}U$ atom produces, until it reaches the final product of $^{206}_{82}Pb$, 8 He nuclei and 6 electrons. A sample of the mineral can be dissolved, melted or pulverized and heated to release the helium, which can then be collected and measured accurately. As we can't be sure that the mineral hasn't lost the gas during its lifetime (for example, due to extreme temperatures), the result of this method indicates an age equal to or less than that of the lead method.

In the same way that He shows the age of uranite, the argon method can be used on potassium-bearing minerals, because the isotope $^{40}_{19}K$ With a half-life of $t = 1.28 \cdot 10^9$ years, it is transformed by radiation β^+ and by K capture (p. 95) into $^{40}_{18}Ar$. Therefore, analysis of the $[^{40}_{19}K : {}^{40}_{18}Ar]$ makes it possible to calculate the age of the potassium mineral. What complicates this method: potassium can decay by an alternative mechanism: it can be transformed under radiation β^- into

$^{40}_{20}Ca$ (see p. 93). This complicates the analysis, as the age determined depends on the percentage of potassium that takes this route.

Another analysis of geological ages is based on the reaction $^{87}_{37}Rb$ $\xrightarrow{4,7\cdot10^{10}\,anos}$ $^{87}_{38}Sr$ $+$ β^- determining the atomic ratio $[^{87}_{37}Rb : ^{87}_{38}Sr]$. The oldest mineral examined by this method was 4.9 billion years old. Minerals containing rhenium (which are rare) can be analyzed using the reaction $^{187}_{75}Re$ $\xrightarrow{4,3\cdot10^{10}\,anos}$ $^{187}_{76}Os$ $+$ β^-.

Energy balance of nuclear reactions

Radiation is mass loss

For now, we've done a mass balance, taking 4 u from the parent isotope during radiation α. We've left the atomic mass unchanged during radiation β and □. When we use the exact masses, however, we will see that this is not quite the case, but there is always a slight discrepancy between the calculated and experimental masses. In reality, we think that some mass is "disappearing".

An example should show this: In the total decay of $^{238}_{92}U$ (exact atomic mass: 238.0508 u) all 8 atoms of $^{4}_{2}He$ (4,0026 u). For the mass of the resulting $^{206}_{82}Pb$ we then calculate a mass of

$$238{,}0508 - (8 \cdot 4{,}0026) = 206{,}0300 \text{ u}$$

But in uranite we found a lead of only 205.9745 u; that makes a difference of

$$206{,}0300 - 205{,}9745 = 0{,}0555 \text{ u}$$

In another context, we have already seen that the radiation β that also occurs during this decay cannot be blamed for this loss of mass, because the electrons emitted are later reabsorbed by the cations in the form of valence electrons. So how do you explain this "disappearance" of 55.5 mg of mass for every mole of uranium?

Of course, we haven't accounted for the energy of this nuclear reaction. More than 50 million electronvolts (eV) are released from each uranium atom, i.e. more than 50 faradayvolts (FV)[22] from each mole of uranium in its transition to lead. The equivalence of mass and energy according to *Einstein* is

$$m = \frac{E}{c^2} \qquad\qquad \text{Equation 10}$$

A mass *m* of 1 g corresponds to an energy *E* of 931.5 million faradayvolts.

Applied to the 55.5 mg of uranium loss, this gives $0.0555 \cdot 931.5 = 51.7$ mi faradayvolts - a value in excellent agreement with the experiment. We can say that $^{238}_{92}U$ undergoes, in its transition to $^{206}_{82}Pb$ not only a <u>material</u> mass loss of 32.0208 u (emission of 8 He), but also an <u>energetic</u> mass loss of 0.0555 u (emission of 51.7 MeV for each atom of^{206} Pb formed).

---

[22] 1 eV· atom^{-1} = 1 FV· mol^{-1} = 96,485 kJ· mol^{-1}

In the radio transition in lead, 35 MeV are involved for each radio atom, due to the emission of radiation α, β and $\square$. According to footnote 22this is 35 MFV (read: mega faradayvolts) or $3 \cdot 10^9$ kJ per mole of radium. The mass equivalent of this energy, according to Equation 10, is 38 mg which "add up" in the decay of 1 mole of Ra.

Cohesive energies in the nucleus

The mass of a nucleus is less than the sum of the masses of its constituents, protons and neutrons. This **"mass defect"** $\square m$ corresponds to the cohesive energy that is released in the construction of the nucleus from the loose constituents; it represents the stability of the atomic nucleus.

To show this, we look at a particularly stable nucleus: helium. As mentioned above, the atomic mass of $_2^4He^{2+}$ is 4.00260 u. The sum of two protons $_1^1H^+$ and two neutrons $_0^1n$ gives

$$(2 \cdot 1{,}007276 + 2 \cdot 1{,}008665) = 4{,}03188u$$

The mass defect is then $\Delta m = 4{,}03188 - 4{,}002600 = 0{,}02928u$; The energy released in the construction of this nucleus is then

$$0{,}02928 \cdot 931{,}5 = 27{,}3 \text{ million faradayvolts}$$

or 2634 million kJ$\cdot$ mol^{-1} , or 662 billion kJ per kg of helium formed.

In equation form we write

$$2\,H^+ + 2\,n \rightarrow He^{2+} + 2{,}634{,}000{,}000 \text{ kJ.} \qquad \text{Equation 11}$$

To break up a helium nucleus, the reverse of the reaction (Equation 11), the incredible energy of 2,634 million kJ$\cdot$ mol^{-1} would be expended. This fact now makes it clear why in the decay α He nuclei are expelled instead of the elementary particles.

Even within nuclear reactions, this value is especially high; the fission of uranium nuclei (p. 101), for example, provides "only" 1/10 of the energy of Equation 11 - when related to the same mass of starting material!

How can the reaction in Equation 11 be carried out? In fact, this reaction is not only extremely exothermic, but also requires an incredibly high activation energy. A fusion of 4 hydrogen nuclei, $_1^1H^+$ forming a He nucleus can only be achieved by heating to several million degrees. This can be achieved, for example, by means of a hydrogen bomb (p. 139) or in the explosion center of a uranium or plutonium bomb. But, of course, this reaction also takes place inside the Sun (p. 18 and p. 106) where it ensures its energy balance.

In order to characterize the stability of nuclei, it is more usual, instead of indicating the mass defect $\Box$m, to refer to the **nuclear cohesive energy** E (unit: MeV) which is given from Equation 10. It has also proved useful to divide these values by the number of nucleons = [protons + neutrons], because this way the elements of the TPE are more comparable to each other. For example, the cohesive energy per nucleon in the case of the nucleus $^4_2He^{2+}$ is 27.3 : 4 = 6.8 MeV. For oxygen, it is calculated from the mass defect of

$$8 \cdot 1{,}00782 \; (^1_1H) + 8 \cdot 1{,}00866 \; (^1_0n) - 15{,}99491 \; (^{16}_8O) = 0{,}1370 \; g$$

a nuclear cohesive energy of

$$0{,}1370 \cdot 931{,}5 = 127{,}6 \; MeV,$$

for the entire nucleus; or 127,6 : 16 = 8,0 MeV by nucleon. For practical purposes, it doesn't matter whether you use only the mass of the nucleus (as in the case of He) or the mass of the whole atom (as in the case of O): in the first case, the mass of the electrons doesn't appear; in the second case, you make a mistake due to the cohesive energy of the electrons, but this energy is very small compared to the nuclear energy. For the 8 electrons arranged in 8 H atoms, $E =$ 0.0001 MeV is observed; in the case of the 8 electrons in the $^{16}_8O$ it is $E = 0.001$ MeV. You can see that the nuclear energy is about 10,000 times greater, so it is permissible to neglect the electrons. A comparison of the stability of the nuclei of all the TPE elements is shown in Figure 14. The higher the position of an element, the greater the stability of its nucleus.

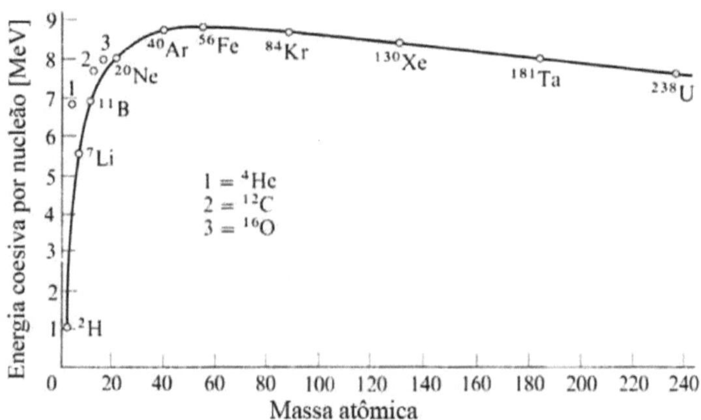

As Figure 14 shows, all the elements, except for a few very light nuclei, have a nuclear cohesive energy of between 7 and 9 MeV (on average: 8 MeV). In the lightest nuclei, the [surface : mass] ratio takes on a high value. As the surface is always a place of higher energy, this value causes the cohesive energy to fall below the average. Thus, for deuterium $^{2}_{1}H$ an energy of only $E = 1.1$ MeV is observed, for tritium $^{3}_{1}H$ of $E = 2.8$ MeV per nucleon.

We can also see another trend in this curve: the heavier elements obviously have less stable nuclei than the medium-mass elements. This is due to the increasingly dominant Coulombian repulsion between protons. Thus, in elements of masses 90 - 140 u, the cohesive energy is on average $E = 8.4$ MeV, while in element 235 it is "only" $E = 7.5$ MeV. Conclusion, to be further refined on p. 101when the $^{235}_{92}U$ nucleus is bombarded by neutrons, it breaks up, preferably into two fragments with masses between 95 and 138 u, respectively. This process is expected to release approximately $235 \cdot (8,4 - 7,5) \approx 200$ MeV.

Maximum nuclear stability, according to Figure 14is in the middle of the curve, near the elements Fe, Co and Ni. This makes it easy to understand why these elements are so abundant in the universe. Don't forget how many of these metals are in the Earth's core...

This fact also provides the explanation that both the fission of heavier nuclei (p. 101) and the fusion of lighter nuclei (p. 106) are exothermic processes. In both cases, the attraction between the nucleons increases and a large amount of cohesive energy is released. Which of these processes is more energetic, fission or fusion? Well, that depends on the reference. Since in the fission of heavy nuclei many nucleons are involved, then the nuclear energy "per mole" is higher. When energy is related to the unit of mass, on the other hand, the fusion of light nuclei is more energetic than fission (p. 106).

We can also see in Figure 14 that $^{4}_{2}He$ deviates from the general line. Likewise the elements with multiples of this mass, $^{12}_{6}C$ e $^{16}_{8}O$ stand out for their extraordinarily high nuclear stability. Equally high is the position of the nucleus $^{8}_{4}B$, which, however, breaks spontaneously ($t_{1/2} = 3 \cdot 10^{-16}$ s) into two nuclei of $^{4}_{2}He$. Once again we recognize the preference for releasing helium nuclei in radioactive processes; it also explains the immense energy that raises the interior of the sun to millions of degrees, due to the nuclear fusion of hydrogen forming helium. And we also have an idea of the devastating effect of a "hydrogen bomb" (p. 139), which "harnesses" the gigantic energy from the fusion of H to He.

Artificial nuclear reactions

Elsewhere (p. 11) the composition of the nucleus was presented and that a certain element is characterized by the number of protons present. The purpose of transforming one element into another is then achieved by changing the number of protons. The most logical way would be to add protons, to produce an element further to the right on the TPE, or else a treatment to lose protons to create an element to the left of the parent element. This treatment could be, for example, bombardment with high kinetic energy particles that have a low positive charge. We think of the neutron ($_0^1 n$; zero charge), a proton ($_1^1 H^+$ charge +1) or the helium nucleus ($_2^4 He^{2+}$; charge +2), because particles with a low positive charge would be less likely to penetrate a nucleus with a high positive charge. In more recent times, however, with the increased power of particle accelerators, the transformation of elements has also been possible by bombardment with nuclei of boron (5 positive charges), carbon (+6), nitrogen (+7), oxygen (+8), fluorine (+9), neon (+10) and even nuclei with a charge above 10 (p. 87).

To date, thousands of nuclear reactions have been carried out; in addition to the 334 natural isotopes, more than 1000 artificial isotopes have been created and we can say that around 2000 different isotopes of around 111 elements are known today. And elementary physics has not yet reached its limits...

Next, we'll look first at single nuclear reactions, where each collision with energetic particles triggers a single elementary act. Then we'll deal with nuclear chain reactions where, as we know from chemical reactions via free radicals, ($H_2 + Cl_2 \rightarrow 2\ HCl$) or ($2\ H_2 + O_2 \rightarrow 2\ H_2 O$), each exothermic elementary act triggers other elementary acts; these mechanisms, when occurring in a well-controlled manner, can be used to continuously produce new elements and energy (nuclear power plant), or, in the case of an uncontrolled event, lead to a new type of explosives and weapons with devastating effects.

Single nuclear reactions

Projectiles and their acceleration

To provoke a reactive collision of nuclei (target nuclei) with $^1_1H^+$ or $^4_2He^{2+}$ the reactants (projectiles) must be accelerated considerably, because only a high-speed collision overcomes the inherent repellency between the particles and makes it possible for the nuclei to penetrate. In the case of particles $^4_2He^{2+}$ we already have an idea of how to produce them, because we know about the natural radioactive substances that emit radiation □: these particles come out of the element with energies of several million eV, so with speeds of a few 10,000 km per second (p. 36).

The history of artificial nuclear reactions really began with the bombardment and breaking up of heavy nuclei using particles α emitted by radioactive substances. Today, however, the attacking particles are no longer restricted to these natural radioactive sources, whose intensity and energy are naturally limited; today artificial radiation α ισ υσεδ, ωιτη high intensity and almost unlimited energies, produced in acceleration chambers. The chamber designed by the American physicist *Ernest Lawrence* (1901 - 1958), where the charged particle describes a spiral trajectory, has proved particularly useful. The equipment is widely used today and is known as a **cyclotron, the** principle of which is shown in Figure 15.

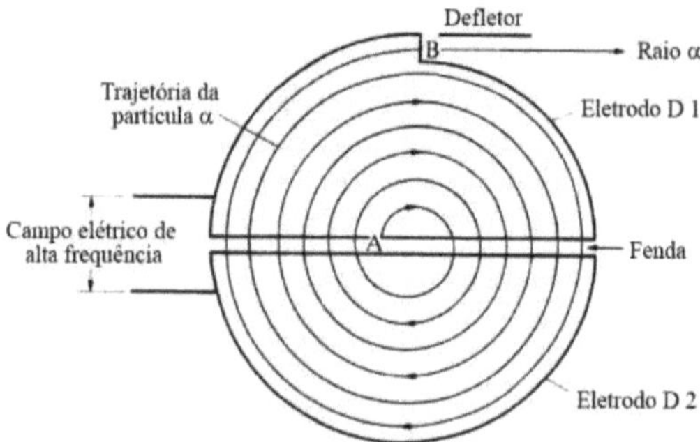

Figure 15 Principle of operation of the cyclotron.

The cyclotron, in its simplest form, consists of two semicircular flat cavities ("D electrodes") in an evacuated discharge chamber. The chamber is located between the poles of a strong magnet (magnetic poles above and below the plane of the paper), operated with a high-frequency alternating voltage. The particles α generated in the space between the two electrodes (location A in Figure 15) are pulled by the electric field into one of the two cavities, where they are, as in a *Faraday* cage, removed from the electric field. Under the influence of the magnetic field, which is homogeneous and directed perpendicularly to the plane of the paper, the particles □□descrevem a semicircle. When they re-enter the space between the two cavities, the particles are affected by the electric field, which changes sign synchronously with the orbital period of the accelerated particle. The ion $^4_2He^{2+}$ undergoes this process several times inside the chamber, while the radius of curvature inside the semicircle is constantly increasing due to the increase in speed (centrifugal force). The particle α μοϖεσ away from the center in a spiral path and, after reaching the desired speed at location B, finally passes through a deflection plate ("deflector") and exits the chamber, heading for its destination.

A "α ray" generated in this way, whose current is 10^{-4} amperes, easy to achieve, has its natural equivalent in 10 kg of pure radium, since 1 g of pure radium (i.e. free of decay products) emits a number of $^4_2He^{2+}$ per second (see p. 40) which corresponds to a current intensity of around 10^{-8} amperes.

The first fully functioning cyclotron had a diameter of 10 cm. Today we are talking about other dimensions: the European research center CERN (*Conseil Européen pour la Recherche Nucléair*) has a proton accelerator, LHC, 200 m in diameter that accelerates particles to 28 billion eV = 28 GeV. The $^1_1H^+$ in this equipment first pass through a "linear accelerator" which pre-accelerates the particles to 50 MeV. Particle accelerators exist today in great variety[23] and number; some important cyclotrons are:

---

[23] Read more at https://pt.wikipedia.org/wiki/Acelerador_de_partículas .
Helmut Wiedemann: Particle Accelerator Physics (= Graduate Texts in Physics). 4th edition. Springer International Publishing, Cham 2015, ISBN 978-3-319-18316-9, doi:10.1007/978-3-319-18317-6 (Open Access).
S Y Lee: Accelerator Physics. 4th edition. World Scientific, 2019, ISBN 978-981-3274-67-9, doi:10.1142/11111 (Open Access book by SCOAP3 program).
Alexander Wu Chao, Maury Tigner, Hans Weise, Frank Zimmermann: Handbook of Accelerator Physics and Engineering. 3rd edition. World Scientific, 2023, ISBN 978-981-12-6917-2, doi:10.1142/13229 (in English).
Ragnar Hellborg (Editor): Electrostatic Accelerators: Fundamentals and Applications (= *Alexander Chao, Christian W. Fabjan, Rolf-Dieter Heuer, Takahiko Kondo, Franceso Ruggiero* [Ed.]: Particle Acceleration and Detection). Springer Berlin Heidelberg, 2005, ISBN 978-3-540-23983-3, doi:10.1007/b138596 (in English).

History of accelerators:
Andrew Sessler, Edmund Wilson: Engines of Discovery: A Century of Particle Accelerators. World Scientific, 2007, ISBN 978-981-270-070-4, doi:10.1142/6272.

- Brookhaven (USA): proton synchrotron at 33 GeV;
- Serpuchow (Russia): proton synchrotron at 70 GeV;
- Weston (USA): 200 GeV;
- Geneva (Switzerland): CERN with 300 GeV; diameter of 2.4 Km, occupying an area of 20 Km^2 ;
- Berkley (USA): Multi-particle accelerator (" Omnitron ") that can accelerate all ions, from H to U, at a wide range of speeds, where heavy ions can reach energies of 500 MeV.

Now let's talk about the other particles, apart from $^4_2He^{2+}$ which serve as projectiles. The simplest and smallest are the proton, neutron and electron.

The **protons** don't need as much acceleration as the He nuclei to penetrate a target nucleus, because they only have half as much positive charge and are therefore less repelled. To bring about a transformation in the target nucleus, they need energies of a few thousand eV - which corresponds to an initial speed of a few thousand kilometers per second. Even with H$^+$ of just a few 10,000 eV, nuclear reactions have been observed - albeit with poor yields. The hydrogen used can be of three types: $^1_1H^+$ (ordinary), $^2_1H^+$ (deuterium) or $^3_1H^+$ (tritium).

The latter two are, due to their high masses, more effective in reactive collision than ordinary hydrogen. Their acceleration usually takes place in the cyclotron.

Neutrons. Naturally, bombardment with protons and particles α becomes more difficult the higher the charge on the target nucleus. Heavy elements are therefore particularly opposed to these projectiles. No such restriction exists with accelerated neutrons. Nuclei of any size and charge can be attacked just as easily using ^1_0n because they have no repelling charge. This means that neutrons don't need much acceleration to provoke a nuclear reaction. Even "slow" neutrons, with energies in the order of 1 eV (speed of a few 10 km per second) can still trigger reactions (see reaction with $^{235}_{92}U$ described in Equation 38).

There are various sources of these neutrons, depending on the intensity (= concentration) required:

- An intensity of 10^4 to 10^7 neutrons per second and cm^2 is achieved with a mixture of radioactive elements with beryllium powder (p. 83). The mixture can be from radiators α ($^{210}_{82}Pb, ^{210}_{84}Po, ^{226}_{88}Ra, ^{228}_{90}Th, ^{239}_{94}Pu, ^{241}_{95}Am$) or radiators γ ($^{124}_{51}Sb$).

- Intensities of 10^8 to 10^{10} neutrons per second and cm^2 can be obtained by attacking deuterium ${}^2_1H$, tritium ${}^3_1H$ or beryllium ${}^9_4Be$ with deuterium nuclei ${}^2_1H^+$ nuclei accelerated in the cyclotron.
- Even higher intensities, from 10^8 to 10^{16} neutrons per second and cm^2 , can be achieved inside the reactor of a uranium-based nuclear power plant (p. 117).

Electrons. Devices similar to the cyclotrons described above have been developed to accelerate electrons (" Betatron s"). Radiators β are sources that can be used to carry out nuclear reactions. Betatrons are also very useful for research into mesons (Table 2). Geneva (Switzerland) has a large electron accelerator (0.6 GeV); Hamburg (Germany) has a larger one (*DESY* = Deutscher Elektronen SYnchrotron; up to 7.5 GeV). Even larger is the linear accelerator in Palo Alto (USA), called SLAC = *Stanford Linear Accelerator*, a straight accelerator 3 km long that takes electrons to an energy of up to 45 GeV.

Kinetic energies. The energy values of these accelerated particles are impressive, but such high energy is not always an advantage, depending on the result you want.

- Low-energy particles, up to a few 10 MeV, enter the target nucleus and remain there; or they launch one or two new elementary particles from the nucleus they hit.
- On the other hand, when the projectiles are quite energetic, some 100 MeV, the result is excessive fragmentation of the nucleus hit, in which large fragments are lost and the target nucleus loses 10 to 50 u in mass.
- Especially interesting is a third category of nuclear reactions, where exactly two fragments are produced from the target nucleus. This is controlled **nuclear fission.** The heaviest and most unstable nuclei require low-energy projectiles to undergo fission. Very slow neutrons are sufficient. On the other hand, when it comes to lighter and more stable target nuclei, the energy of the projectile must be much higher.
- A fourth category of nuclear reactions, caused by high-energy projectiles, is **nuclear fusion**. Unlike fission, this causes the formation of heavier nuclei from light target nuclei.

Here is a description of these simple nuclear reactions, with examples and applications.

Simple nuclear reactions

This introductory text is limited to the most commonly used "projectiles", which are helium nuclei, hydrogen nuclei, neutrons and radiation □. We will only briefly mention that there is also bombardment with heavier nuclei, such as B, C, N, O, F and Ne, which produces a series of new elements and isotopes, including the heaviest (radioactive, short half-life) elements in TPE.

Nuclear reactions with helium nuclei

When a particle α with high kinetic energy hits a target nucleus, it can be integrated into that nucleus. Example: $^{7}_{3}Li + ^{4}_{2}He \rightarrow ^{11}_{5}B$. But in most cases this collision is accompanied by the expulsion of an elementary particle from the target, a proton or a neutron.

Proton emission. When a proton is ejected during bombardment with $^{4}_{2}He^{2+}$ the nucleus of charge k and mass m is transformed into a new element of nuclear charge $(k+1)$ and mass $(m+3)$:

$$^{m}_{k}E + ^{4}_{2}He \rightarrow ^{m+3}_{k+1}E + ^{1}_{1}H \qquad\qquad \textit{Equation 12}$$

The first test with this type of reaction dates back to 1919, a milestone in the history of radioactivity. The English physicist *Lord Rutherford* succeeded in creating a new chemical element for the first time. He let the very energetic α particles (6 MeV), generated by $^{212}_{83}Bi$ (= thorium C), into a nitrogen atmosphere. On a fluorescent screen just behind the collision site, he observed intense scintillations caused by the particles α (range: up to 7 cm) that had not collided. But in addition to these, he also noted some faint scintillations whose range was greater (up to 40 cm). Through an exact mathematical analysis, he was able to prove that the latter were generated by protons. His bold conclusion: these $^{1}_{1}H^{+}$ were ejected from the affected N nuclei. Subsequent investigations confirmed this assumption and proved that some nitrogen turned into oxygen. At this point we must admire the human ingenuity that was able to deduce, from the flash of a few dots of light on a fluorescent screen, the solution to an ancient mystery and unrealizable dream of mankind, the artificial transformation of the elements.

According to the general equation, Equation 12, the oxygen formed in *Rutherford*'s experiment has a mass of 17 u:

$$^{14}_{7}N + ^{4}_{2}He \rightarrow ^{17}_{8}O + ^{1}_{1}H \qquad\qquad \textit{Equation 13}$$

This type of reaction is called the " process $\square$, p", the reaction in Equation 13, in particular, can be written as

$^{14}_{7}N\ (\alpha,\text{p})\ ^{17}_{8}O$,

where the first symbol in brackets represents the projectile ($\alpha = {}^{4}_{2}He^{2+}$) and secondly the emitted particle (p = ${}^{1}_{1}H^{+}$). This notation will be used next to describe simple nuclear reactions.

Unlike most of the radioactive reactions presented so far, the process in Equation 15 is endothermic. This is verified when analyzing the exact atomic masses involved in this reaction:

16,999130 (${}^{17}_{8}O$) + 1,007825 (${}^{1}_{1}H$) - 14,003074 (${}^{14}_{7}N$) - 4,002603 (${}^{4}_{2}He$) = +0,001278 g

The increase in mass corresponds to an energy of

$0.001278 \cdot 931.5 = 1.19$ MeV.

Therefore, this reaction requires particles α of sufficiently high kinetic energy. We can bet that in the first instance an excited fluorine atom is formed, ${}^{18}_{9}F$ *which then breaks down to form ${}^{17}_{8}O$. Extensive experiments in this direction have revealed that reaction 16 is accompanied by a parallel reaction, where the intermediate nucleus ${}^{18}_{9}F$ * stabilizes itself by releasing a neutron. The parallel product is then ${}^{17}_{9}F$. The relative amounts of main and parallel product depend on the energy delivered by the particle $\square$.

The flight of the particle α in reaction 13 and the formation of two new nuclei can be visualized by our eye using a " **Wilson chamber** ", a closed box containing supersaturated steam. The chamber is first saturated with water vapor or alcohol. A sudden expansion causes the temperature in this chamber to drop (*Joule-Thomson* effect) and causes the state of supersaturation. This is when the nuclear reaction takes place. The atomic debris collides with the gas molecules and generates ions which, in turn, represent embryos for the formation of liquid droplets (= fog). Using advantageous lighting and a dark background, this mist is visible in the form of trails, as we know from jet planes in the blue sky. Photographs of this situation look like the sketches in Figure 16. The formation of the fog is especially effective in the case of particles $\square$: in the radiation α, β and γ ιν comparison, the ionization probabilities of the atmosphere in the *Wilson* chamber behave like 100,000 : 100 : 1. Therefore, the Wilson chamber is more suitable for visualizing the trajectories of particles $\square$.

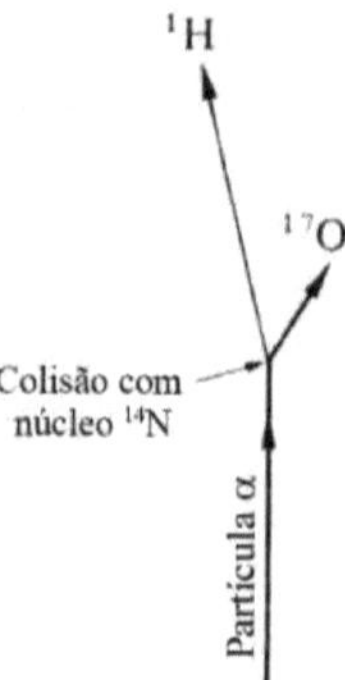

Figure 16 Schematic of a nuclear reaction caused by a particle ,
visualized in the *Wilson* chamber.

Few particle trajectories α end suddenly and split into two new trajectories in different directions. A thin trail caused by the proton and a stronger trail by the oxygen nucleus. An analysis of the moments involved in the collision reveals that the masses of the fragments are actually 1 and 17 respectively.

Let's not forget that a transformation of nuclei according to reaction 13 is a very rare event - especially when using natural α radioactive sources. Of approximately 100,000 particles α, only one collides favorably with a nitrogen atom. This shows the great experimental difficulty of proving the products H_2 and O_2 from this reaction. To get a sense of the orders of magnitude:

1 g of radium generates 167 mm^3 of helium in one year due to radiation □. With 100% shock efficiency, one molecule of H_2 and one of O_2 would be produced for every two atoms of He. Over the course of a year, that would be around 80 mm^3 of H_2 and 80 mm^3 of O_2 . However, as the efficiency is very low, only 80 : 100,000 = 0.0008 mm^3 of these gases will be produced, i.e. less than 1/1000 mm^3 ! The Austrian chemist *F. A. Paneth* developed methods for detecting such small quantities of gas. With these, for example, it was possible to determine the age of minerals and meteorites using the helium method (p. 62) with precision. However, by taking advantage of the radiation α from the cyclotron, which is often more intense than from a natural source, the rate of reactive collisions can be multiplied. In this way, it would be possible to produce gases in suitable volumes and times.

Similar to the reaction $^{14}_{7}N$ (α,p) $^{17}_{8}O$ it is possible to transform lithium into beryllium, boron into carbon, fluorine into neon, sodium into magnesium, aluminum into silicon, silicon into phosphorus, phosphorus into sulfur or calcium into scandium.

Emission of neutrons. Bombarding nuclei with particles α can also cause neutrons to be ejected instead of protons. In this case, a new element of nuclear charge (*k+2*) and mass (*m+3*) is generated from an element *E* of nuclear charge k and mass *m:*

$$^m_k E \;+\; ^4_2 He \;\rightarrow\; ^{m+3}_{k+2} E \;+\; ^1_0 n \qquad\qquad \text{Equation 14}$$

An important reaction in this category is the transformation of He and Be nuclei into neutrons and carbon:

$$^9_4 Be \;+\; ^4_2 He \;\rightarrow\; ^{12}_6 C \;+\; ^1_0 n \qquad\qquad \text{Equation 15}$$

This reaction still serves today as an effective source of neutrons in laboratories ("neutron cannon") which, in turn, can be used for other nuclear reactions (p. 83). In practice, a closed glass ampoule containing salts of radium ($^{226}_{88} Ra$) or americium ($^{241}_{95} Am$) as radiation sources α, mixed with powdered beryllium metal. Alternatives are metal alloys of Po/Be, Pu/Be or Am/Be, trapped in stainless steel containers. The neutrons formed have high kinetic energy: in the case of radium they have up to 7.8 MeV, i.e. a speed of 39,000 Km s^{-1} . As they have no charge, neutrons easily pass through the wall of the container and can come into contact with matter that touches the container from the outside. The neutron yield, however, is lower than in radiation sources □: around 30 neutrons correspond to one million particles □. High neutron intensities, as already mentioned on p. 73is achieved by accelerating particles in the cyclotron. Other sources of neutrons, see p. 81 e 117.

In 1930, German physicists *W. Bothe* and *H. Becker* discovered neutrons through the reaction in Equation 15. However, they misinterpreted the high penetrating power of the particles in matter, as they were betting on a form of radiation □. Only a year later, in 1931, the *Joliot-Curie* couple showed that this radiation was able to expel high-energy protons when it hit solid paraffin. Finally, the English physicist *J. Chadwick* proved in 1932, in his work in *Rutherford*'s laboratory, that it really wasn't massless radiation, but neutral particles with a relative mass of 1 u (exact value: 1.008665012 u; see Table 6) - those that *Rutherford* himself had already predicted 12 years earlier and *W.D. Harkins* in 1921 named "neutrons", with the symbol "n". Free neutrons are radioactive and decay in a half-life of 10.6 minutes. The products are radiation β^- and protons:

$$\ _0^1 n \ \rightarrow \ \ _1^1 p^+ \ + \ \ _{-1}^0 e^- \ + \ \bar{\nu} \qquad\qquad \text{Equation 16}$$

As its nuclear charge is zero, the neutron should appear in the TPE before hydrogen. But it's also true that it has no electrons in its sphere, so it doesn't make any chemical reactions or bonds, so in most TPE representations it is omitted.

Within the general scheme of Equation 14, it is also possible to transform:

Li in B, B in N, C in O, N in F, F in Na, Na in Al, Mg in Si, Al in P, Si in S, P in Cl and K in Sc.

In total, well over 100 of these processes are known today ($\square$, n). If the He nuclei have sufficient kinetic energy, it is also possible to expel more than one neutron from the target nucleus (to date it has been possible to expel 9 neutrons at once). More than 60 processes ($\square$, 2n) and more than 60 processes ($\square$, 3n) are known.

Although heavy target nuclei repel projectiles more $_2^4 He^{2+}$ more than light ones, it has already been possible to transform some of these using very high-energy α radiation (in the order of 300 MeV):

$$\ _{33}^{75} As \ + \ \ _2^4 He \ \rightarrow \ \ _{35}^{78} Br \ + \ \ _0^1 n$$

Nuclear reactions with protons

As already mentioned on p. 72to achieve a nuclear reaction, protons don't require as high kinetic energies as the He nucleus, because they only have a positive charge and are less repelled by the target nucleus. Lower energy also means that it is more likely that the proton will simply be absorbed, without any nucleons being expelled from the target nucleus. That doesn't mean it's impossible. By accelerating the protons sufficiently, He nuclei, protons and neutrons can be expelled, as will be exemplified below.

Proton capture. The simple incorporation of H^+ into the target nucleus increases the nuclear charge by 1, with the simultaneous emission of a quantum of $\square$. The product is the neighboring element to the right in the TPE:

$$_kE + {}_1H \rightarrow {}_{k+1}E + {}_0h\nu \qquad\qquad \text{Equation 17}$$

The mass of the new element $_{k+1}E$ increases by 1 or 2 - depending on whether the bombardment occurs with protons or deuterons:

$$^m_kE + {}^1_1H \rightarrow {}^{m+1}_{k+1}E + {}_0h\nu \quad \text{or} \quad {}^m_kE + {}^2_1H \rightarrow {}^{m+2}_{k+1}E + {}_0h\nu.$$

Through this mechanism it is possible to transform Li into Be, Be into B, C into N, F into Ne or Si into P. Today > 25 of these processes are known (p, □).

Emission of particles □. If the bombardment of nuclei with protons leads to the expulsion of a particle □, the result is the reverse reaction of Equation 12:

$$_kE + {}_1H \rightarrow {}_{k-1}E + {}_2He \qquad\qquad \text{Equation 18}$$

Again, the mass of the new element, located to the left of the starting element, depends on the type of hydrogen used for the bombardment,[1] H or H:[2]

$$^m_kE + {}^1_1H \rightarrow {}^{m-3}_{k-1}E + {}^4_2He \quad \text{or} \quad {}^m_kE + {}^2_1H \rightarrow {}^{m-2}_{k-1}E + {}^4_2He.$$

An example of this mechanism (p, □) is the conversion of lithium into helium. Obviously an easy reaction, due to the stability of the He nucleus, which is much higher than Li (compare Figure 14):

$$^7_3Li + {}^1_1H \rightarrow {}^4_2He + {}^4_2He \qquad\qquad \text{Equation 19}$$

Equation 19 was the first nuclear reaction carried out with artificially accelerated projectiles (*Cockroft, Walton* 1932). Nor does it serve to produce helium in remarkable volumes, as will be shown in the following estimate:

When protons of energy 0.2 MeV are applied in reaction 24, only one out of 100 million protons can penetrate a Li nucleus. This, after all, comes as no surprise when we compare the size of the nucleus (radius in the order of femtometers = 10^{-15} m) with the size of the entire Li atom. The probability of hitting a nucleus is like hitting a bean in a room 1000 m^3 - without aiming! If you attack lithium metal with a proton current of 1 mA (that's about the maximum intensity that today's devices can achieve) for a whole year, the nuclear reaction would produce not much more than 1 mm^3 of He gas. This shows that this process (p, $\square$) certainly <u>does not</u> serve to produce new elements in useful quantities, in all analogy to the processes that use He nuclei as projectiles (p. 74).

Incidentally, the same applies to the energy aspects of the nuclear reactions presented so far. Let's assume that reaction 24 is exothermic, with a remarkable 17.3 MeV, in the form of high-kinetic-energy He nuclei. This is the result calculated from the mass loss of 0.001863 g· mol^{-1} in this reaction, after subtracting the 0.2 MeV of energy invested in accelerating the protons. Just the fact that we have to accelerate 100 million H$^+$ to achieve just one reactive collision makes it clear that we are still wasting millions of times more energy than we are producing:

$$\Delta H = (10^8 \cdot 0{,}2) - 17{,}3 = +2{,}7 \ \text{MeV}$$

In contrast to these reactions, the bombardment of heavy nuclei (starting with thorium) using neutrons revolutionized the world's energy economy, as will be discussed on p. 114. 112. Because the mechanism of these reactions is not in single steps, but in a chain.

Incidentally, the same result as in reaction 19 is obtained when attacking light lithium nuclei, $_3^6Li$ with deuterons:

$$_3^6Li \ + \ _1^2H \ \rightarrow \ _2^4He \ + \ _2^4He \qquad\qquad Equation \ 20$$

And one more reaction (p, $\square$): reaction 21 produces tritium, the super-heavy and radioactive hydrogen, which we'll talk about later:

$$_4^9Be \ + \ _1^2H \ \rightarrow \ 2_2^4He \ + \ _1^3H. \qquad\qquad Equation \ 21$$

What's interesting about reaction 21 is that the particles α emitted here have higher energies, 11 MeV per particle, than any natural source of radiation $\square$. However, in today's cyclotrons it is possible to produce He nuclei with energies 3 to 4 times higher!

Other examples of (p, $\square$) and (d, $\square$) processes are Be → Li, B → Be, C → B, N → C, F → O, Na → Ne, Mg → Na, Al → Mg, Si → Al, Fe → Mn. The total number of reactions studied exceeds 50.

Emission of protons. Firing nuclei with protons, so that protons come out, naturally doesn't change the mass or the element. You might think.

$$_k E \;+\; _1 H \;\rightarrow\; _k E \;+\; _1 H' \qquad\qquad\qquad \textit{Equation 22}$$

However, this type of reaction can generate other isotopes if the H-projectile and the H emitted have different masses. For example, if$_k$ E elements are bombarded with deuterons or tritons and ordinary protons are emitted, this leads to the isotopes of elements with$_{k+1}$ E and$_{k+2}$ E, respectively:

$$_k^m E \;+\; _1^2 H \;\rightarrow\; _k^{m+1} E \;+\; _1^1 H \;,\; _k^m E \;+\; _1^3 H \;\rightarrow\; _k^{m+2} E \;+\; _1^1 H$$

A particularly interesting case of this type of nuclear reaction is in the case of

$$_1^2 H \;+\; _1^2 H \;\rightarrow\; _1^3 H \;+\; _1^1 H \;+\; 4 \text{ MeV} \qquad\qquad \textit{Equation 23}$$

This reaction first produced hydrogen of mass 3 u ("Tritium"; symbol "t" or "T"; M.*L.E. Oliphant, P. Harteck, E. Rutherford* 1934).

Tritium is a radioactive isotope ($t_{1/2}$ = 12,346 years) that transforms into $_2^3 He$ under radiation emission β^- :

$$_1^3 H \;\rightarrow\; _2^3 He \;+\; _{-1}^0 e$$

Similarly, Li, Be, B, C, N, Na and Al can be transformed into their heavier isotopes. Today more than 160 processes (d, p) and around 15 processes (t, p) are known. (t, p).

Emission of neutrons. Bombarding nuclei with protons can cause a neutron to be emitted; the new element is then located to the right of the parent nucleus:

$$_k E \;+\; _1 H \;\rightarrow\; _{k+1} E \;+\; _0 n \qquad\qquad\qquad \textit{Equation 24}$$

Depending on the application of protons, deuterons or tritons, the mass of the new element is the same, 1 u more or 2 u more than the starting element:

$$^{m}_{k}E \;+\; ^{1}_{1}H \;\rightarrow\; ^{m}_{k+1}E \;+\; ^{1}_{0}n \;;$$

$$^{m}_{k}E \;+\; ^{2}_{1}H \;\rightarrow\; ^{m+1}_{k+1}E \;+\; ^{1}_{0}n \;;$$

$$^{m}_{k}E \;+\; ^{3}_{1}H \;\rightarrow\; ^{m+2}_{k+1}E \;+\; ^{1}_{0}n$$

It is not uncommon for two or more neutrons to be ejected (up to 14). Around 120 (p, 2n) and (d, 2n) processes are known today; more than 70 (p, 3n) and (d, 3n) processes.

A particularly interesting reaction of the type in Equation 24 is the attack of deuterium with deuterons. As an alternative to reaction 27, it can yield a helium atom of mass 3 u and a neutron:

$$^{2}_{1}H \;+\; ^{2}_{1}H \;\rightarrow\; ^{3}_{2}H \;+\; ^{1}_{0}n \;+\; 3{,}2 \text{ MeV} \qquad\qquad Equation\ 25$$

Helium of this type would be ideal for an airship or weather balloon, as it is non-flammable, inert and 25% lighter than ordinary helium. However, as has already been explained several times (p. 74, p. 79), the yield of this reaction is so small that the preparation of ^{3}He in such quantities is unimaginable. On the other hand, reaction 25 does serve as an effective source of neutrons. Similarly to reaction 15, it is independent of the presence of radioactive substances, so it can serve as a neutron cannon, as discussed on p. 75. 73. This method can generate very high neutron intensities in a well-adjusted cyclotron - equivalent to a mixture of 100 kg of beryllium emanation. The energy of the neutrons is adjustable; with the use of sufficiently energetic deuterons, the neutrons come out with an energy of up to 20 MeV.

There are two alternatives to reaction 25: tritium attack (T = $^{3}_{1}H$; adsorbed on titanium or zirconium) or beryllium metal ($^{9}_{4}Be$) also generates neutrons in high concentration:

$$^{3}_{1}H \;+\; ^{2}_{1}H \;\rightarrow\; ^{4}_{2}He \;+\; ^{1}_{0}n \;;\; ^{9}_{4}Be \;+\; ^{2}_{1}H \;\rightarrow\; ^{10}_{5}B \;+\; ^{1}_{0}n$$

Other reactions according to Equation 24 are the transmutation of lithium into beryllium, boron into carbon, carbon into nitrogen, nitrogen into oxygen, oxygen into fluorine, fluorine into neon, sodium into magnesium or aluminum into silicon. In short, hundreds of (p, n) and (d, n) reactions are known. In the case of heavy target nuclei and, at the same time, high kinetic energy protons, nucleus fission can occur. An example of this is the attack of copper with protons of 50 to 60 MeV, where the fragments are chlorine and aluminum:

$$^{65}_{29}Cu \ + \ ^{1}_{1}H \ \rightarrow \ ^{38}_{17}Cl \ + \ ^{27}_{13}Al \ + \ ^{1}_{0}n$$

Nuclear reactions with neutrons

In the simplest case, bombarding a nucleus with neutrons leads to its absorption. Of course, the element doesn't change, but a heavier isotope is created. In addition, a quantum γ ισ being emitted:

$$^{m}_{k}E \ + \ ^{1}_{0}n \ \rightarrow \ ^{m+1}_{k}E \ + \ ^{0}_{0}h\nu \qquad\qquad\qquad Equation\ 26$$

This type of reaction is now known for almost all TPE elements; hundreds of reactions (n, □) have been documented. As neutrons are not repelled by nuclear charge and also to prevent nucleons from being ejected, the neutrons for this reaction must be especially <u>slow</u>. So how do you generate neutrons with low kinetic energy? From most sources they are emitted at unacceptably high energies. If they are conducted through a stretch of water or solid paraffin, they lose kinetic energy as they undergo elastic collisions with hydrogen, contained in these media in high concentration.

The neutron capture reaction, Equation 26, is especially important with the heavier target elements, as we will see on p. 101. But it is equally applicable to light elements, with the aim of generating new isotopes of them. Examples are:

$$^{23}_{11}Na \rightarrow ^{24}_{11}Na, \ ^{27}_{13}Al \rightarrow ^{28}_{13}Al \ \ \text{or} \ \ ^{63}_{29}Cu \rightarrow ^{64}_{29}Cu.$$

Naturally, these isotopes are not separable from their parent element, as they have identical chemical behavior. Generally, they are radiators β^- , the most common case with isotopes with an excess of neutrons (because a neutron turns into a proton). The important "neutron activation analysis" takes advantage of this fact.

With *Prompt gamma neutron activation analysis* (PGAA[24]), trace elements can be detected in mixtures such as minerals, metal alloys or other samples. The element in question is attacked by slow neutrons and transforms into its radioactive isotope, which can be identified after moving away from the neutron source. Most of the time, the radiation γ (rather than the radiation $\square$) of the element to be identified is used; the half-life and intensity allow conclusions to be drawn as to the type and concentration of the respective element.

The pioneering work was done by *E. Fermi,* who showed in 1936 that a series of elements became radioactive after being irradiated with neutrons. In the same year, *G.v. Hevesy demonstrated* traces of Dy (0.01%) in a preparation of Y and traces of Eu in a sample of Gd, by irradiating them with a natural neutron source (p. 73), a mixture of radium and beryllium.

Other examples where neutron activation analysis has been successfully applied:

$^{55}_{25}Mn$ ($t_{1/2}$ = 2.58 h); $^{76}_{33}As$ ($t_{1/2}$ = 26.4 h); $^{198}_{79}Au$ ($t_{1/2}$ = 2.685 d); they all emit radiation β and $\square$. The sensitivity of the analysis is so high that quantities of just 10^{-10} g can still be detected.

A neutron activation analysis that attracted media attention was carried out in 1961 on a special sample of just a few milligrams: a lock of *Napoleon I*'s hair. The sample dated from the day after the emperor's death (05/05/1821) on the island of St. Helens. It was cut off and kept separately. The analysis led to the conclusion that he did not die a natural death, but was the victim of arsenic poisoning. Not only the presence and quantity of As could be proven: an analysis, section by section of the 13 cm strands, a length that corresponds to the growth within the last year of his life, revealed that the poison was applied at irregular intervals; even the dates when this happened were estimated...

When higher energy neutrons are applied in the bombardment, they cause neutrons to be expelled from the target nucleus. The result is lighter isotopes, if $\geq$ 2 neutrons are expelled. An example is the reaction $^{100}_{42}Mo \rightarrow {}^{99}_{42}Mo$ where the process is (n, 2n). Dozens of processes of this type are known. The projectile neutron must have an energy of at least 8 MeV, because the cohesive energy of neutrons in most target nuclei (see Figure 14) is around 8 MeV.

For the expulsion of three neutrons, the process (n, 3n), you then need projectile neutrons of at least 24 - 8 = 16 MeV.

---

[24]   https://www.iaea.org/resources/databases/database-for-prompt-gamma-ray-neutron-activation-analysis

Emission of protons. When nuclei are bombarded with high-energy neutrons, protons can be lost. This process represents the reverse reaction of Equation 24, the result being the element to the left of the hand element:

$$^{m}_{k}E \; + \; ^{1}_{0}n \; \rightarrow \; ^{m}_{k\text{-}1}E \; + \; ^{1}_{1}H \qquad\qquad\qquad \textit{Equation 27}$$

A particularly important example of this type of nuclear reaction is the transformation of ^{14} N $\rightarrow^{14}$ C. This process also occurs in nature, as will be seen on p. 98:

$$^{14}_{7}N \; + \; ^{1}_{0}n \; \rightarrow \; ^{14}_{6}C \; + \; ^{1}_{1}H$$

Interestingly, the product here, ^{14} C, is unstable and decays under radiation emission β^{-}. It then turns back into N:^{14}

$$^{14}_{6}C \; \rightarrow \; ^{14}_{7}N \; + \; ^{0}_{-1}e$$

The isotope ^{14} C is used as a radioactive indicator to investigate the mechanisms of organic reactions. Equally important is its use in determining the age of plant and animal organisms, as will be shown on p. 98.

Similarly, it is possible to transform:

F $\rightarrow$ O; Na $\rightarrow$ Ne; Al $\rightarrow$ Mg; S $\rightarrow$ P; Cr $\rightarrow$ V; Fe $\rightarrow$ Mn; Ni $\rightarrow$Co; Zn $\rightarrow$ Cu; Pd $\rightarrow$ Rh, among others.

Around 100 such processes (n, p) are known today.

The reactive events of this reaction are, as we already know, quite rare (1 out of every 10^{8} neutrons hits the target and reacts). Even rarer are the events where neutrons expel, instead of a proton (p = $^{1}_{1}H$), a triton (t = $^{3}_{1}H$) from the target nucleus. To achieve this, the projectiles must be very fast. There are cosmic processes where such neutrons are generated - as well as radiation γ which acts as "activation energy". They can interact with the nitrogen in the Earth's atmosphere, according to

$$^{14}_{7}N \; + \; ^{1}_{0}n \; \rightarrow \; ^{3}_{1}H \; + \; ^{12}_{6}C$$

One of the products is tritium, which in turn is a radiator β^- that turns into[3] He. This is the most widely accepted explanation for the existence (in small quantities) of tritium and[3] He in our atmosphere (compare p. 98 and footnote 28).

Emission of particles $\square$. The expulsion of helium nuclei by highly energetic neutrons is the reverse reaction of Equation 14; the product is the element two places to the left in the TPE:

$$^{m}_{k}E \ + \ + \ ^{1}_{0}n \ \rightarrow \ ^{m-3}_{k-2}E \ + \ ^{4}_{2}He \qquad\qquad \text{Equation 28}$$

This type of reaction is used to transform lithium into tritium, a reaction that occurs in remarkable quantity in the nuclear reactor (p. 115):

$$^{6}_{3}Li \ + \ + \ ^{1}_{0}n \ \rightarrow \ ^{3}_{1}H \ + \ ^{4}_{2}He \qquad\qquad \text{Equation 29}$$

The two products are gases, T_2 and He. Tritium is adsorbed on uranium in the form of UT_3 . This chemisorption is relatively strong, but reversible: when the uranium compound is heated to 500 °C, the tritium is desorbed. If you want to separate the gases T_2 and He, the chemical reaction of tritium to $T_2 O$ (= liquid) is offered. This makes separation very easy.

Similar are the transformations

B → Li; Al → Na; P → Al; Cl → P; Sc → K; Mn → V; Co → Mn; Ge → Zn; Th → Ra; among others.

More than 50 such processes (n, $\square$) are known.

As neutrons pass through the atoms of a gas without ionizing them, they do not cause traces in the *Wilson* chamber (p. 75). They are only evident from their reactive collision where they produce two ions. The fragments formed in this reaction are visible, as outlined in Figure 17Suddenly two trails form from one point. The trajectory of the neutron, however, remains invisible.

Figure 17 Schematic of a typical *Wilson* chamber image of a neutron reaction.

To sum up, we can say that slow neutrons favor process 26, medium-speed neutrons favor process 27 and fast neutrons favor process 28. The fission of heavy nuclei caused by neutrons is, due to its high economic importance, preserved for the chapter "Control of nuclear chain fission".

Nuclear transformations with heavier projectiles

So far we have discussed nuclear reactions where the target element has shifted up to two places in the TPE, either to the left or to the right. These transformations can be visualized briefly, as shown in the beryllium example in Figure 18. We have adopted the designation of the processes as introduced on p. 74on the abscissa the number of protons, on the ordinate the sum [p + n].

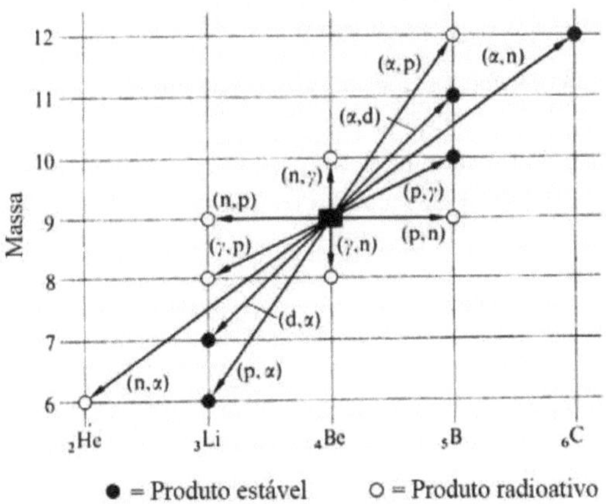

Figure 18 Beryllium's "nuclear transformation spider".

Considerably greater changes in the nucleus are achieved when attacking with projectiles heavier than $_1$H or $_2$He, namely with the nuclei of

$$^{6}_{3}Li,\ ^{9}_{4}Be,\ ^{11}_{5}B,\ ^{12}_{6}C,\ ^{14}_{7}N,\ ^{16}_{8}O,\ ^{19}_{9}F,\ ^{20}_{10}Ne.$$

Bombardment with these projectiles is hampered by the multiple positive charges that collide, so the kinetic energy must be considerably high. On the other hand, this usually results in the expulsion of nucleons; it has been shown that almost exclusively neutrons are emitted. Therefore, the nuclear charge of the target element is increased by 3, 4, 5, 6, 7, 8, 9 and 10, respectively.

A few examples of these reactions are listed in the following table.

$$^{58}_{28}Ni + ^{6}_{3}Li \rightarrow ^{1}_{0}n + ^{63}_{31}Ga$$
$$^{12}_{6}C + ^{11}_{5}B \rightarrow 3\,^{1}_{0}n + ^{20}_{11}Na$$
$$^{65}_{29}Cu + ^{12}_{6}C \rightarrow 3\,^{1}_{0}n + ^{74}_{35}Br$$
$$^{75}_{33}As + ^{12}_{6}C \rightarrow 5\,^{1}_{0}n + ^{82}_{39}Y$$
$$^{79}_{35}Br + ^{12}_{6}C \rightarrow 3\,^{1}_{0}n + ^{88}_{41}Nb$$
$$^{115}_{49}In + ^{22}_{6}C \rightarrow 4\,^{1}_{0}n + ^{123}_{53}Cs$$
$$^{121}_{51}Sb + ^{12}_{6}C \rightarrow 7\,^{1}_{0}n + ^{126}_{57}La$$
$$^{181}_{73}Ta + ^{12}_{6}C \rightarrow 5\,^{1}_{0}n + ^{188}_{79}Au$$
$$^{197}_{79}Au + ^{12}_{6}C \rightarrow 9\,^{1}_{0}n + ^{200}_{85}At$$
$$^{203}_{81}Tl + ^{12}_{6}C \rightarrow 5\,^{1}_{0}n + ^{210}_{87}Fr$$
$$^{206}_{82}Pb + ^{12}_{6}C \rightarrow 5\,^{1}_{0}n + ^{213}_{88}Ra$$

$$^{115}_{49}In + ^{14}_{7}N \rightarrow 4\,^{1}_{0}n + ^{125}_{56}Ba$$
$$^{141}_{59}Pr + ^{14}_{7}N \rightarrow 6\,^{1}_{0}n + ^{149}_{66}Dy$$
$$^{165}_{67}Ho + ^{14}_{7}N \rightarrow 5\,^{1}_{0}n + ^{174}_{74}W$$
$$^{182}_{74}W + ^{14}_{7}N \rightarrow 5\,^{1}_{0}n + ^{191}_{81}Tl$$
$$^{197}_{79}Au + ^{14}_{7}N \rightarrow 6\,^{1}_{0}n + ^{205}_{86}Rn$$

$$^{65}_{29}Cu + ^{16}_{8}O \rightarrow 2\,^{1}_{0}n + ^{79}_{37}Rb$$
$$^{96}_{44}Ru + ^{16}_{8}O \rightarrow 5\,^{1}_{0}n + ^{107}_{52}Te$$
$$^{115}_{49}In + ^{16}_{8}O \rightarrow 5\,^{1}_{0}n + ^{126}_{57}La$$
$$^{139}_{57}La + ^{16}_{8}O \rightarrow 6\,^{1}_{0}n + ^{149}_{65}Tb$$
$$^{140}_{58}Ce + ^{16}_{8}O \rightarrow 6\,^{1}_{0}n + ^{150}_{66}Dy$$
$$^{141}_{59}Pr + ^{16}_{8}O \rightarrow 7\,^{1}_{0}n + ^{150}_{67}Ho$$

$$^{142}_{60}Nd + ^{16}_{8}O \rightarrow 6\,^{1}_{0}n + ^{152}_{68}Er$$
$$^{144}_{62}Sm + ^{16}_{8}O \rightarrow 6\,^{1}_{0}n + ^{154}_{70}Yb$$
$$^{181}_{73}Ta + ^{16}_{8}O \rightarrow 5\,^{1}_{0}n + ^{192}_{81}Tl$$
$$^{197}_{79}Au + ^{16}_{8}O \rightarrow 9\,^{1}_{0}n + ^{204}_{87}Fr$$

$$^{141}_{59}Pr + ^{19}_{9}F \rightarrow 8\,^{1}_{0}n + ^{152}_{68}Er$$
$$^{142}_{60}Nd + ^{19}_{9}F \rightarrow 8\,^{1}_{0}n + ^{153}_{69}Tm$$
$$^{144}_{62}Sm + ^{19}_{9}F \rightarrow 8\,^{1}_{0}n + ^{155}_{71}Lu$$

$$^{140}_{58}Ce + ^{20}_{10}Ne \rightarrow 8\,^{1}_{0}n + ^{152}_{68}Er$$
$$^{141}_{59}Pr + ^{20}_{10}Ne \rightarrow 8\,^{1}_{0}n + ^{153}_{69}Tm$$
$$^{144}_{62}Sm + ^{20}_{10}Ne \rightarrow 7\,^{1}_{0}n + ^{157}_{72}Hf$$
$$^{142}_{60}Nd + ^{20}_{10}Ne \rightarrow 8\,^{1}_{0}n + ^{154}_{70}Yb$$

There are many of these reactions, usually carried out in the universal cyclotron " Omnitron" (p. 72); today we know about 100 reactions with $_6$C, more than 40 reactions with $_7$N and more than 70 reactions with O. $_8$

These reactions are especially important in the synthesis of heavier elements, nuclei that are far beyond the natural ones. This is how the elements $_{98}$Cf (Californium) to $_{106}$Sg (Seaborgium) were produced. In this context, we should also mention the reactions with exotic projectiles with which the heaviest nuclei have been synthesized, for the time being:

Target core	Projectile	Product
Bi	Cr	$_{107}$Bo (Borium; manganese group)
Bi	Fe	$_{108}$Ha (Cassium; iron group)
Pb	Fe	$_{109}$Mt (Meitnerium; cobalt group)

Nuclear reactions induced by rays γ

Perhaps we shouldn't be surprised, given the high energy of this radiation: γ radiation, which has a fairly short wavelength, can also cause a reaction in the nucleus. In analogy to *Lenard*'s legendary essay ("Photoelectric effect", 1902) which studied the interaction of photons with the electronic layer, this process here can be called the **photo nuclear effect**. When we look back a little, we realize that the reactive impact of γ radiation on a nucleus represents the reverse reaction of the events abstracted by Equations 17 and 26. Moreover, γ radiation can easily lift a nucleus into high energy states - which can be used to examine solid matter (see indication for *Mössbauer* spectroscopy in footnote 11).

We can refer in this context to the breaking down of deuterons into protons and neutrons:

$$\mathrm{^{2}_{1}H} \;+\; \mathrm{^{0}_{0}}h\nu \;\rightarrow\; \mathrm{^{1}_{1}H} \;+\; \mathrm{^{1}_{0}}n.$$

One of the best-known reactions in this category is the irradiation of beryllium powder $\mathrm{^{9}_{4}Be}$:

$$\mathrm{^{9}_{4}Be} \;+\; \mathrm{^{0}_{0}}h\nu \;\rightarrow\; 2\,\mathrm{^{4}_{2}He} \;+\; \mathrm{^{1}_{0}}n$$

In this reaction, of course, very stable nuclei are formed. This explains the high quantum yield and opens up the possibility of using mixtures of radiators γ with Be as neutron sources (compare p. 73). Especially advantageous is the mixture with $\mathrm{^{124}_{51}Sb}$ ($t_{1/2} = 60.3$ days) which emits γ rays of up to 2.09 MeV; alternatives are mixtures of beryllium powder or metal alloys with $\mathrm{^{24}_{11}Na}$, $\mathrm{^{88}_{39}Y}$, $\mathrm{^{116}_{49}In}$ or $\mathrm{^{140}_{57}La}$.

With high-energy γ rays, in the order of 20 MeV, reactions can even be achieved with particularly stable nuclei, such as those of silver:

$$\mathrm{^{107}_{47}Ag} \;+\; \mathrm{^{0}_{0}}h\nu \;\rightarrow\; \mathrm{^{106}_{47}Ag} \;+\; \mathrm{^{1}_{0}}n$$

Around 50 processes ($\square$, n) and ($\square$, p) met in the 1990s, but that list must have grown a lot...

Another curiosity is the reaction of boron with very high-energy γ radiation (up to 250 MeV), which led to a rather exotic nucleus:

$$^{11}_{5}B + ^{0}_{0}h\nu \rightarrow 3\,^{1}_{1}H + ^{8}_{2}He$$

As you can imagine, a He nucleus of mass 8 is not at all stable. It decays ($t_{1/2} =$ 122 milliseconds) under radiation emission β in lithium which, in turn, is a radiator β and finally decays providing two nuclei of ordinary helium:

$$^{8}_{2}He \rightarrow ^{8}_{3}Li + ^{0}_{-1}e \rightarrow 2\,^{4}_{2}He + ^{0}_{-1}e_{\cdot}$$

Again: the decay of unstable nuclei into quite advantageous fragments such as $^{4}_{2}He$ is not surprising from an energy point of view. Other reactions of this type (they are not processes that require radiation $\square$):

$$^{9}_{3}Li \rightarrow 2\,^{4}_{2}He + ^{1}_{0}n + ^{0}_{-1}e$$

$$^{8}_{4}Be \rightarrow 2\,^{4}_{2}He$$

$$^{8}_{5}B \rightarrow 2\,^{4}_{2}He + ^{0}_{+1}e$$

$$^{12}_{7}N \rightarrow 3\,^{4}_{2}He + ^{0}_{+1}e$$

The missing elements from the periodic table

Until 1937, TPE still had white spots: places 43 (Tc; technetium), 61 (Pm; promethium), 85 (At; astatine) and 87 (Fr; francium) were empty. The reason is that there are no stable isotopes of these elements; they are all radioactive. So, over billions of years, they practically disappeared from the Earth's crust. However, through the nuclear reactions presented in the last few chapters, these nuclei can be created. Thus, the first syntheses date back to

1937Tc

1945Pm

1940At

1939Fr   .

The chemical properties of these elements are well known today.

Technetium. Italian researchers *C. Perrier* and *E. Segrè* (1937) irradiated molybdenum with deuterons to produce technetium, an element that is no longer found in nature. To date, 21 isotopes of this element have been made - almost all of them short-lived. Thus, element 43 exists with masses 90 to 110 u. There are also two nuclide isomers of each of the following isomers: 90, 91, 93, 94, 95, 96, 97, 99 and 102. In short, there are 30 different nuclei of technetium. Some of these, with their most important characteristics, are listed below; metastable nuclei are marked with *; above the reaction arrow the main decay process and the half-life time (s = seconds; m = minutes; d = days; a = years).

$$^{93}_{43}\text{Tc}^* \xrightarrow{\gamma;\,43{,}5\text{ m}} {}^{93}_{43}\text{Tc} \qquad {}^{96}_{43}\text{Tc}^* \xrightarrow{\gamma;\,52\text{ m}} {}^{96}_{43}\text{Tc} \qquad {}^{98}_{43}\text{Tc} \xrightarrow{\beta^-;\,4{,}2\cdot10^6\text{ a}} {}^{98}_{44}\text{Ru}$$

$$^{93}_{43}\text{Tc} \xrightarrow{\beta^+;\,2{,}7\text{ h}} {}^{93}_{42}\text{Mo} \qquad {}^{96}_{43}\text{Tc} \xrightarrow{K;\,4{,}3\text{ d}} {}^{96}_{42}\text{Mo} \qquad {}^{99}_{43}\text{Tc} \xrightarrow{\beta^-;\,2{,}1\cdot10^5\text{ a}} {}^{99}_{44}\text{Ru}$$

$$^{95}_{43}\text{Tc}^* \xrightarrow{K;\,60\text{ d}} {}^{95}_{42}\text{Mo} \qquad {}^{97}_{43}\text{Tc}^* \xrightarrow{\gamma;\,91\text{ d}} {}^{97}_{43}\text{Tc} \qquad {}^{100}_{43}\text{Tc} \xrightarrow{\beta^-;\,15{,}8\text{ s}} {}^{100}_{44}\text{Ru}$$

$$^{95}_{43}\text{Tc} \xrightarrow{K;\,20\text{ h}} {}^{95}_{42}\text{Mo} \qquad {}^{97}_{43}\text{Tc} \xrightarrow{K;\,2{,}6\cdot10^6\text{ a}} {}^{97}_{42}\text{Mo} \qquad {}^{101}_{43}\text{Tc} \xrightarrow{\beta^-;\,14{,}2\text{ m}} {}$$

We noticed the following trends: the low-mass isotopes decay preferentially under positron emission or K capture (p. 95) forming molybdenum; the high-mass isotopes decay preferentially under negatron emission forming ruthenium.

Only three isomers have significant half-lives:

$^{97}_{43}Tc$ ($t_{1/2}$ = 2,6· 10^6 years), $^{98}_{43}Tc$ ($t_{1/2}$ = 4.2· 10^6 years), $^{99}_{43}Tc$ ($t_{1/2}$ = 2.12· 10^5 years).

Of these, only $^{99}_{43}Tc$ is produced in relevant quantities in the uranium nuclear reactor, where it is formed in good yields, with more than 6.2% Tc in the fission products (Figure 20). In a reactor with a power of 100 MW (= 100,000 kJ s^{-1} ; this is a value that is neither too large nor too small for today's plants), around 4 g of Tc are produced per day, i.e. 1.5 g of Tc. $^{99}_{43}Tc$ per day, or 1.5 kg per year. This, together with its half-life of 212,000 years, allows it to be marketed like any other stable element. With the quantity and capacity of nuclear power plants that exist today, it would be easy to produce more Tc than there is of its heavier counterpart, (stable) rhenium, in the entire earth's crust...

Another possible source of the element Tc is the treatment of the neighboring nucleus, $_{42}$ Mo, with neutrons, protons, deuterons or particles □.

Its isolation involves an oxidation step where pertechnetate, TcO_4^- , is formed. This can be separated from the U and Pt compounds and then extracted into an organic phase (methylpyridines), from where it is isolated by vapor drag.

Promethium. As Pm occurs in nature in only tiny quantities ($< 10^{-19}$ %; the rarest of the lanthanides) it must also be made artificially. It was discovered in 1945 by the Americans *J.A. Marinsky, L.E. Glendenin* and *C.D. Coryell,* as one of the fragments in the fission of uranium. O $^{147}_{61}Pm$ has a half-life of $t_{1/2} = 2.62$ years. This isotope is also the most important because only it can be produced in sufficient quantities to study its chemical properties. In the uranium nuclear reactor, as described above, Pm is produced with a yield of 2.6% - which makes it possible to isolate up to ½ Kg over the course of a year. It is a radiator β that emits relatively soft radiation (only 0.223 MeV).

Alternative syntheses include bombarding the neighboring element, $_{60}$ Nd, with protons, deuterons or α particles. It is also possible to irradiate $_{60}$ Nd with high-intensity neutrons in a nuclear reactor:

$$^{146}_{60}\text{Nd} \xrightarrow{+n; -\gamma} {}^{147}_{60}\text{Nd} \xrightarrow{-\beta^-} {}^{147}_{61}\text{Pm} \quad .$$

There are now 28 known isotopes of promethium, all of which are radioactive. They are either transformed into neodymium by radiation β^+ or K capture, or into samarium by radiation β^- . The most stable isotope is $^{145}_{61}Pm$ with $t_{1/2} = 17.7$ years. Promethium is a typical lanthanide, with chemical behavior that lies between its neighbors, Nd and Sm. The pure element is a fairly non-noble metal (ε^0 (Pm/Pm^{3+}) = -2.423 V), whose oxidizability is comparable to magnesium.

Astat. Astat is the heaviest halogen, and its occurrence in nature is almost nil (only $3 \cdot 10^{-24}$ % of the earth's crust). At occurs as a short-lived intermediate of the three natural decay series (Table 3), its parent element is polonium in the form of $^{215}_{84}Po, {}^{216}_{84}Po$ e $^{218}_{84}Po$. However, its weight in the radioactive balance (p. 55) is extremely small, for example, for every $7 \cdot 10^{16}$ atoms of $^{238}_{92}U$ atoms, only one atom of $^{218}_{85}At$.

The atom must then be made artificially, the most common way being to bombard Bi with particles α accelerated in the cyclotron to an energy of ~30 MeV:

$$^{209}_{83}\text{Bi} + {}^{4}_{2}\text{He} \rightarrow {}^{211}_{85}\text{At} + 2\,{}^{1}_{0}n$$

To date, 24 isotopes of At are known, from masses 196 to 219 u. Half-life times vary between 10^{-7} seconds and 8.3 hours. As At does not appear as a fragment of uranium fission, there is little chance of producing this element in notable quantities.

Francium. The element with the number 87 is the heaviest alkali metal. All the isotopes of Fr - 30 are known today - are very short-lived. It was therefore only in 1939 that the French researcher (hence the name of the element) *M. Perey* discovered the isotope of mass 223 u, which is an intermediate in the natural decay series of actinium.

The most stable isotope (it's better to say the least unstable) is $^{223}_{87}Fr$ which decays with $t_{1/2} = 21.8$ minutes, one part producing the halogen$_{85}$ At and radiation $\square$, the other part the alkaline earth metal$_{88}$ Ra and radiation $\square$.

Its occurrence as an intermediate in the natural decay series is so minuscule that small quantities of this element must be obtained artificially. The most common way is to bombard $^{226}_{88}Ra$ with neutrons from the uranium reactor. This provides actinium in the first instance, $^{227}_{89}Ac$ However, due to the instability of all isotopes, it has not been possible to produce notable quantities of francium to date.

Artificial radioactivity and its application to natural events

In previous chapters, we learned about nuclear reactions triggered by the bombardment of nuclei (target nuclei) by accelerated particles (projectiles), including helium nuclei (p. 74), hydrogen nuclei (p. 78), neutrons (p. 83), heavier nuclei (p. 87), as well as rays γ (p. 89). Most of the time, a new nucleus is formed which is not stable, but radioactive; the product decays to form more stable successor nuclei, following a first-order kinetic law (p. 47). These cases can be called **artificial radioactivity**. In addition to providing an overview of these reactions, this chapter also broadens the range of mechanisms by which an unstable nucleus can stabilize itself. It also provides evidence that unstable nuclei, although artificial, can be generated by nature itself and, finally, how to use their decay kinetics in the service of science.

The first studied case of artificial radioactivity dates back to 1934, when the couple *I. Curie* (daughter of *Marie Curie*; 1897 - 1956) and *F. Joliot* (1900 - 1958) attacked a piece of aluminum with radiation α from polonium. They observed and proved, using chemical methods, that new elements had formed:

A large part of the new elements was stable silicon (95%), which was formed according to the reaction

$$\ _{13}^{27}Al \ + \ _{2}^{4}He \ \rightarrow \ _{14}^{30}Si \ + \ _{1}^{1}H;$$

But an unstable phosphorus was also formed to a lesser extent (5%). Relatively quickly, with $t_{1/2} = 2.50$ minutes, it $_{15}^{30}P$ decays into stable silicon. A new form of radiation was observed in this decay: radiation β^+, i.e. positrons:

$$\ _{13}^{27}Al \ + \ _{2}^{4}He \ \rightarrow \ _{15}^{30}P \ + \ _{0}^{1}n$$

$$\ _{15}^{30}P \ \xrightarrow{\ 2,50 \ min\ } \ _{14}^{30}Si \ + \ _{1}^{0}e^+$$

It's remarkable how it was proven at the time that positron radiation really was generated by phosphorus. There were two pieces of chemical evidence:

1) When the piece of aluminum was dissolved in hydrochloric acid (Al + 3 $H^+ \rightarrow Al^{3+}$ + 3 "H"), the radioactivity was not washed into the solution, but volatilized in the form of phosphane (P + 3 "H" $\rightarrow PH_3 \uparrow$).
2) When the irradiated aluminum was dissolved in aqua regia (strongly oxidizing; 2 P + 2.5 "O" + 3 $H_2O \rightarrow 2\ H_3PO_4$), adding a little phosphate and zirconium salt, the radioactivity was quantitatively conducted to the precipitate, which in this case was zirconium phosphate; nothing remained in solution.

What was new at the time: positron radiation, $_{1}^{0}e^+$ whose energy can be calculated from the mass loss in the nuclear reaction and the mass-energy equivalent of the positron (0.51 MeV). For example, in the reaction

$$\ _{29}^{64}Cu \ \rightarrow \ _{28}^{64}Ni^- \ + \ _{1}^{0}e^+$$

a mass defect of

$63{,}92922$ ($_{29}^{64}Cu$) - $63{,}92797$ ($_{28}^{64}Ni^-$) = 0.00125 g·mol^{-1}, which corresponds to 1.17 MeV of energy per atom. Hence the kinetic energy of the positrons is as follows

$$1.17 - 0.51 = 0.66 \text{ MeV.}$$

This value was actually observed in the experiment.

Since the pioneering work of *Curie-Joliot* (Nobel Prize in Chemistry in 1935), countless cases of radioactivity have been observed and characterized. Well over 1,000 artificial isotopes have been added to the 81 naturally occurring elements. Today, 111 elements are known, each of which has at least one radioactive isotope; most elements have several - not to say many - radioactive isotopes.

Most unstable isotopes decay under emission of either positive or negative electrons. The emission of positrons from the nucleus generates new neutrons at the cost of protons. For the element, this means a shift in the TPE to the left.

There is a mechanism equivalent to the emission of positrons: the integration of negatrons into the unstable nucleus. This can actually occur, as *L.W. Alvarez* discovered in 1937, as the atom or ion takes advantage of an electron closer to the nucleus. As the first electronic layer is the K layer, the process was called **capture** K.

K-capture can be easily detected when we remember how an X-ray tube works (which is nothing more than electromagnetic radiation released when an electron from an outer shell fills a vacancy in an inner shell. There are still selection rules that restrict the number of possibilities for [origin → destination] in these electron drops. Thus, filling a hole in the K layer gives rise to K radiation$_\alpha$ or K$_\beta$, when the electron fills the vacancy in the L layer it is called L radiation$_I$, L , L$_{IIIII}$, etc. [25]

An interesting detail: K-capture is the only radioactive process in which the element's electronic layer is involved, and this is the only form of decay that varies according to the element's chemical environment. This means that the observed phenomena involving K-capture depend (slightly) on the chemical composition of the sample. Therefore, the half-life of $^{7}_{4}Be$ in a sample of BeF_2 is around 0.08% longer than in a sample of Be metal.

One more word on the equivalence of positron radiation and K capture.

During K-capture, the nuclear mass increases, not much, but undeniably, due to the mass of an electron (m_e - = 0.000 548 580 26 u) <u>and</u> the mass equivalent to the energy of this exothermic reaction (0.783 MeV). The reaction can be represented as

$$p^+ \;+\; e^- \;+\; 0{,}783\,\text{MeV} \;\rightarrow\; n$$

---

[25] For details, see "selection rules" and "X-ray spectrum" in a physical chemistry textbook, for example: *G. Wedler*, Handbook of Chemical Physics, Calouste Gulbenkian **2001**.

In the other case, the emission of positrons, the mass of the nucleus decreases by the mass of the positron itself (m_{e+} = 0.000 548 580 26 u), but at the same time increases by the mass equivalent of the energy of 1.805 MeV (endothermic!). This value, incidentally, corresponds to the mass of the twins [electron + positron] that are formed in an annihilation (p. 14), adding 0.783 MeV from the exotherm of the annihilation. The energy balance of the reaction

$$p^+ \; + \; 1{,}805 \, \text{MeV} \; \rightarrow \; n \; + \; e^+$$

reveals that the increase in mass is the same: in both mechanisms [m_e - + 0.783 MeV]. This means that the new isotope has exactly the same mass - no matter by which of the two competing mechanisms it was formed.

Finally, through both processes, either K capture or emission of β^+, the attacked element is transformed into another, just to its left in the TPE. An example:

The natural isotope $^{40}_{19}K$ (Table 4) is partially transformed (11%) to the stable natural isotope of $^{40}_{18}Ar$ (footnote 9), by means of the K capture mechanism. The remaining 89% is transformed into $^{40}_{20}Ca$ under the emission of radiation β ($^-$ Table 4).

An emission of He nuclei, as observed with naturally decaying elements (Table 3), is almost exclusively observed in heavy nuclei, but rarely in light nuclei (exception: $^8_4Be \; \rightarrow \; ^4_2He \; + \; ^4_2He$); on the other hand, decay under positron emission can only be observed in artificially unstable isotopes.

One might ask: what is the criterion for emitting a negatron, β^-, as has been mentioned many times in this book, and when emitting a positron, β^+ ? The answer lies in the properties of the nucleus resulting from the transformation, in other words, its ratio [protons : neutrons]. This ratio can be non-ideal (outside the black line in Figure 11), or very large or very small. Hence:

> ➤ Radioactive isotopes where the proportion of neutrons is quite high. These are, for example, those that have absorbed neutrons (p. 83). They are always likely to emit negatrons, β^-, because through this process a neutron can be transformed into a proton, according to
>
> $$n \rightarrow p^+ + e^- \uparrow$$
>
> Thus, the [p : n] ratio can be corrected. This reaction is usually exothermic.

➢ On the other hand, those elements that were made by proton absorption (p. 78) show a tendency to emit positrons, because in the reaction

$$p^+ \rightarrow n + e^+ \uparrow$$

the [p : n] ratio can be corrected downwards. This reaction is usually endothermic.

The speed of any of these emissions strictly follows the first-order law, as explained on p. 47.

Artificial radioactive elements are very welcome, as they greatly expand the stock of elements that could serve as indicators (English: *tracer*). The concentration of an element during a chemical reaction or biological process can be monitored with a sensitivity that no other method can achieve: radioactive labeling. The emission of radiation makes it possible to trace an isotope and thus the respective element as a whole, in some reagent or product of a reaction, whether *in vitro* or *in vivo*. The particles emitted by this isotope can be identified and quantified easily and with high sensitivity. Isotopes are particularly valuable in this regard:

Radioactive isotope	Issue	Half-life time
$^{3}_{1}H$ (tritium)	β^-	12,346 years
$^{14}_{6}C$ (compare reaction mentioned on p. 85)	β^-	5,760 years
$^{24}_{11}Na$	β^-	15.03 hours
$^{32}_{15}P$	β^-	14.22 days
$^{35}_{16}S$	β^-	87.5 days
$^{45}_{20}Ca$	β^-	163 days
$^{59}_{26}Fe$	β^-	44.6 days
$^{65}_{30}Zn$	β^-	244 days
$^{89}_{38}Sr$	β^-	50.5 days
$^{131}_{53}I$	β^-	8.04 days

Through these *tracers* it is possible to follow the path of a certain substance (reagent, medicine, organism) and study even quite complex processes, which affect the following areas of research:

- Chemistry (mainly organic chemistry[26] , including intramolecular rearrangements)
- Analytical (trace elements, forensic research)
- Biochemistry (how enzymes work, for example)
- Physiology (metabolic reactions)
- Medicine and pharmacy (drug action, pharmacokinetics)
- Agriculture (absorption in plants, action of bacteria)

Let's stay for a moment with our main element of life: carbon. The isotope[14] C is a radiator β^- and can be used to determine the age (better said: the death) of historical and prehistoric organisms. The method is also known as " **Clock carbon**". Under the influence of hard cosmic radiation, as already explained on p. 85nitrogen in the atmosphere can be transformed into carbon. Through this nuclear reaction, an equilibrium has been established over millions of years in the form of [$^{14}CO_2$:$^{12}CO_2$] contained in the air[27] . In this equilibrium, 16 decays of[14] C occur for every gram of carbon and every minute. The concentration at[14] C is therefore quite small. However, it is sufficient to be analyzed with high sensitivity. The same concentration is found in plants that assimilate carbon dioxide from the air and also in animals that feed on plants. Throughout their lives they are in perpetual contact with the atmosphere, so they incorporate the same relative concentration of[14] C as is present in the atmosphere.

This changes when the organism dies. Hence the incorporation of CO_2 and thus the consumption of[14] C stops. Within 5760 years the concentration of the isotope[14] C drops by half, at 11,520 years after death the activity has reduced to ¼, and so on. This fact allows us to calculate the year when the organism still had 100% activity, i.e. the year it died. This could be a plank from a sunken ship, a bone from a prehistoric animal - any material that still contains carbon - whether in organic (wood) or inorganic (carbonate; hydroxyapatite) form. From there, the carbon must be isolated by thermal / chemical methods (remember that radioactive decay is not noticeably influenced by either the chemical

---

[26] Regarding their usefulness in inorganic chemistry, we recommend starting with the pioneering review article: *D.R. Stranks, R.G. Wilkins.* Isotopic tracer investigations of mechanism and structure in inorganic chemistry. Chem.Rev. **57** (1957), 743-866.

[27] Before the Industrial Revolution there was 0.03% CO_2 in the air; today that rate has risen to 0.04%.

environment or the temperature; compare p. 49). The pioneer of this application was *W.F. Libby* (1947), who facilitated experimental verification of historical and prehistoric dates. With the "carbon clock" it is possible to indicate the date of death, between 400 and 30,000 years ago, with an accuracy of ±5%. One example: the age of a plank from the burial ship of *King Sesostris III of* Egypt (1887 - 1849 BC) confirmed the age indicated by archaeologists, 3,800 years.

Not only the isotope 14 of carbon that was formed according to the

$$^{14}_{7}N \ + \ ^{1}_{0}n \ \xrightarrow{\gamma} \ ^{14}_{6}C \ + \ ^{1}_{1}H$$

is valuable in analyzing archaeological ages, but also the radioactive isotope of hydrogen, tritium, $^{3}_{1}H$ which was formed in a parallel nuclear reaction with nitrogen in the atmosphere:

$$^{14}_{7}N \ + \ ^{1}_{0}n \ \xrightarrow{\gamma} \ ^{12}_{6}C \ + \ ^{3}_{1}H \qquad \text{(Parallel}$$

reaction)

$^{3}_{1}H$ is a ραδιαδορ β⁻ , so it can be used as a *tracer* to determine the ages of aquatic organisms[28] . Over millions of years, an equilibrium has been established where approximately 1 atom of T is present for every 10^{17} atoms of ordinary hydrogen, $^{1}_{1}H$.

By cutting off the free exchange of water with the environment, the concentration of tritium decreases, with a half-life of 12,346 years. Therefore, the analysis of the current tritium concentration can serve to prove the age at which the water (or the analyte containing mainly water) was isolated from the environment. Groundwater can be analyzed, where the result reveals whether it is recent rainwater or whether it really is ancient, isolated groundwater. A particularly eye-catching analysis using the "tritium watch" is the identification of valuable wines and aged drinks (up to 50 years old)...

In addition to the applications described above, the isotopes of artificial radioactivity are constantly gaining importance in the field of medicine, mainly diagnostic, but less therapeutic. They are easy to dose and relatively inexpensive. And most importantly: compared to natural radioactive elements,

---

[28] Fun fact: tritium is even present in our atmosphere. 10 cm^3 of air contains on average one atom of T. But that's honestly too little to be used as a *tracer* - given that the entire atmosphere (!) of the earth contains only 6 g of T.

they can remain inside the body without causing adverse reactions - since their decay products are all light and harmless elements. These "radionuclides" include:

> $^{24}_{11}Na$, decreasing with $t_{1/2}$ = 15.03 hours to $^{24}_{12}Mg$ according to: $^{23}_{11}Na \rightarrow {}^{24}_{12}Mg + {}^{0}_{-1}e + \gamma$.
> ^{24}Na is produced by bombarding ordinary sodium, $^{23}_{11}Na$, with neutrons;
> $^{32}_{15}P$, a phosphorus obtained from ordinary phosphorus, $^{31}_{15}P$ by bombardment with neutrons. It decays to ordinary sulfur, $^{32}_{16}S$, with a half-life of 14.22 days, according to: $^{32}_{15}P \rightarrow {}^{32}_{16}S + {}^{0}_{-1}e + \gamma$.

And finally, these artificially produced isotopes can serve as "batteries" in satellites where long and reliable activity is expected.

The use of these isotopes in various cutting-edge technological fields today is due to their accessibility: they are produced, not only on a gram scale, but in kilograms, in uranium-based nuclear power plants (p. 115).

Total fragmentation of the nucleus

A much greater impact than in the simple nuclear reactions presented so far are very high-energy projectiles, of the order of a few 100 MeV. The bombardment of any TPE element leads to a wide variety of radioactive debris, whose order number extends over 10 - 20 units (new elements) and whose mass differs from the parent nucleus by 20 - 50 units. A few examples should illustrate the devastating effect of these projectiles:

The bombardment of $^{75}_{33}As$ with particles α of 400 MeV produces, among other things, the radioactive isotope $^{38}_{17}Cl$ ($t_{1/2}$ = 37.18 minutes); this fragment has 16 fewer protons and 37 u less mass than the fired nucleus.

The irradiation of $^{63/65}_{29}Cu$ with 200 MeV deuterons or 400 MeV α particles gives a wide spectrum of new elements and isotopes. Only in the manganese fraction have all the isotopes from 51 to 56 been found.

The irradiation of $_{26}Fe$ with 340 MeV protons found many radioactive fragments of the elements sodium (e.g. $^{22}_{11}Na$ e $^{24}_{11}Na$) to cobalt (e.g, $^{55}_{27}Co$ e $^{56}_{27}Co$).

Bombarding aluminum with protons of 1 - 3 GeV leads to almost complete disintegration, roughly described as

$$^{27}_{13}Al + {}^{1}_{1}H \rightarrow 10\,{}^{1}_{1}H + 11\,{}^{1}_{0}n + {}^{7}_{4}Be$$

All that's left are protons, neutrons and beryllium.

This last example shows that very high-energy projectiles cause the nucleus to break up completely, and the fragments can be protons and neutrons - possibly in the form of lighter nuclei.

Fragmentation is especially easy in heavy nuclei, which are already less stable (Figure 14). Thus, deuterons of 50 MeV are enough to break up uranium $^{235/238}_{92}U$ producing lighter isotopes that differ by up to 10 protons and 20 to 30 mass units from the parent nucleus, for example, the $^{211}_{85}At$. It's easy to imagine that the variety of debris becomes excessive when these heavy nuclei are bombarded with high-energy projectiles. Let's take the example of uranium: an attack with particles α oϕ 400 MeV produces isotopes of <u>all</u> the TPE elements with numbers between 25 and 92. In this dismantling, we can bet that the heavier fragments, $Z > 70$ and $M > 180$ approximately, were generated by fragmentations described in the recent examples, while the lighter fragments ($Z \approx 46 \pm 20$; $M \approx 120 \pm 60$) were generated, more probably, by breaking the nucleus of$_{92}$ U into two fragments, a mechanism that will be discussed in detail in the next chapter.

Similar effects that the particles mentioned above have γ rays of very high energy, generated in the "betatron" (p. 73). For example, the irradiation of $^{28}_{14}Si$ with electromagnetic radiation of 100 MeV (corresponding to a wavelength $\lambda \approx$ 1/100 pm) transforms it into sodium ($^{24}_{11}Na$) - which implies the emission of 3 protons and 1 neutron.

Controlled fission of nuclei

In 1938, the German chemists *Otto Hahn* and *Fritz Strassmann* discovered a particularly important nuclear reaction that later led to the "Atomic Age". In their pioneering experiments, they irradiated heavy nuclei with low-energy neutrons. The result of the nuclear reaction is the break-up of the parent nucleus into exactly two unequal fragments. This reaction was called **nuclear fission**.

Let's start with an extremely important reaction: the irradiation of uranium with slow neutrons. The nucleus $^{235}_{92}U$ absorbs the neutron and forms the isotope $^{236}_{92}U$. This, in turn, is quite unstable, nothing more than an intermediate, in English literature described as a "*compound*" nucleus. It spontaneously undergoes fission under the emission of a gigantic amount of heat, 15 billion kJ = 160 million faradayvolts per mole of isotope. $^{236}_{92}U$ This corresponds to 65 billion kJ per kilogram of uranium.

Four years before this discovery, the Italian physicist *Enrico Fermi* had already performed nuclear fission. Only he interpreted the fragments as transuranic elements. If he had correctly interpreted the fission he made, as *Hahn* and *Strassmann* did, we can postulate that the development of the nuclear power plant and the atomic bomb would have been brought forward by four years. Not a few historians believe that this could have prevented the Second World War, since from the discovery of the nuclear reaction to its "use" as a devastating weapon really didn't take long. In any case, these events show how the course of world politics depends on the cognitive powers of researchers...

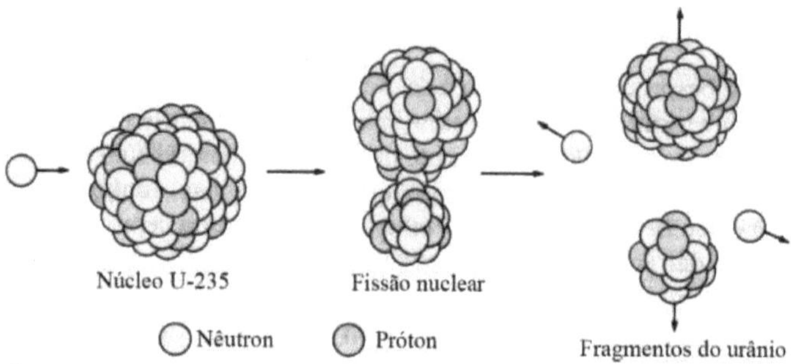

Figure 19   Fission of the^{235} U nucleus into two fragments by bombardment with slow neutrons.

As already mentioned, the fragments formed in this fission usually have different masses. From uranium, the most commonly observed masses are around M = 95 and M = 138 u. Therefore, typical pairs are Kr / Ba; Sr / Xe; Y / I; Br / La. In any case, the sum of the order numbers of each pair is $Z = 92$. The distribution of the new elements is shown in Figure 20. As mentioned above, the elements with masses around 95 and 138 u are especially frequent (with yields of over 6% each), but moving away from these ideal masses the yields drop rapidly (since the ordinate is on a logarithmic scale). In Figure 20 shows the atomic % of the fragments. As each uranium nucleus yields two fragments, the integral under the entire curve gives 200%.

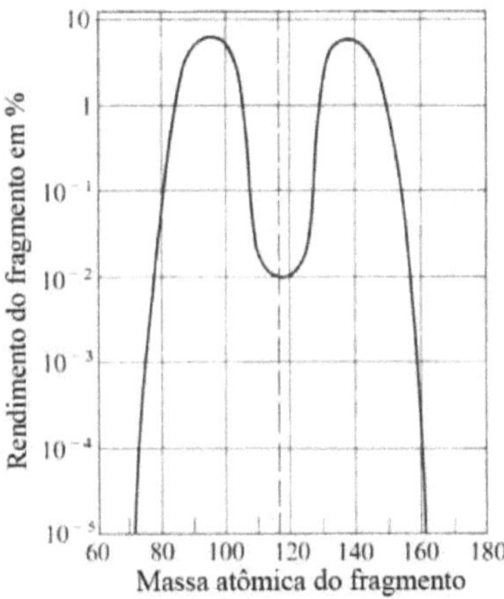

Figure 20 The fragments of uranium nuclear fission and their yields.

At the same time, neutrons are released - an average of 2 to 3 neutrons for each elementary act.

The new elements usually have an excess of neutrons in their nuclei, i.e. they are isotopes located above the black line ("stability line") of the Figure 11. Therefore, they are radiators β^-, a process that transforms neutrons into protons and generates new neighboring elements whose isotopes are closer to the stability line. This is why the curve in Figure 20 refers to the fragments formed directly during fission, not the elements formed after their subsequent decay. However, entire decay series can be observed in the fission reactor, for example:

$$^{92}_{36}\text{Kr} \xrightarrow{\beta^-;\ 1,84\,s} {}^{92}_{37}\text{Rb} \xrightarrow{\beta^-;\ 4,5\,s} {}^{92}_{38}\text{Sr} \xrightarrow{\beta^-;\ 2,71\,h} {}^{92}_{39}Y$$

$$\xrightarrow{\beta^-;\ 3,54\,h} {}^{92}_{37}Zr \ (\text{estável})$$

37 different elements were identified, from $^{72}_{30}Zn$ to $^{161}_{66}Dy$ with M masses from 72 to 161 u, represented by a total of 300 isotopes. Most of the isotopes are radioactive, but 80 of them are stable isotopes, representing the ends of ~30 decay series. In addition to a lot of energy, the uranium fission reaction today serves to produce rare elements and isotopes on a gram and kilogram scale, including:

$^{90}_{38}Sr$; $^{99}_{43}Tc$; $^{137}_{55}Cs$; $^{147}_{61}Pm$.

To the energy values we must also add the heat released in these decay series. All the reactions together in a fission reactor therefore release around 200 million faradayvolts = 19 billion kJ for every mole of uranium (p. 117). This figure far exceeds any other nuclear reaction studied to date. And one more advantage of this fission: as 2 to 3 new energetic neutrons are formed in each elementary act, these can be used to continue the fission, a typical chain reaction kinetics. It can be used for both good and bad, as we'll see below.

O $^{235}_{92}U$ is not the only nucleus that makes this reaction remarkable. The other natural isotope of uranium, o $^{238}_{92}U$ is much more abundant (99.3%). Its irradiation with slow neutrons leads to their absorption. The product is the isotope $^{239}_{92}U$ which is very unstable and decomposes under radiation β⁻ giving the transuranic element, neptunium Np.

$$^{239}_{92}U \xrightarrow{\beta^-} {}^{239}_{93}Np$$

As we already know about the actinide elements, they are all unstable. So neptunium decomposes to form plutonium $^{239}_{94}Pu$ (one of the most toxic elements, by the way) which in turn is radioactive. Free decay then continues.

$$^{239}_{93}Np \xrightarrow{\beta^-} {}^{239}_{94}Pu$$

Many other heavy nuclei can be fissioned by neutrons, including the isotopes $^{233}_{92}U$, $^{238}_{92}U$, $^{239}_{94}Pu$, $^{241}_{94}Pu$, $^{242}_{95}Am$. But so far only the fissions of $^{235}_{92}U$, $^{239}_{94}Pu$ e $^{233}_{92}U$ (in order of decreasing political and economic importance). These, at the same time, are the nuclei where fission is induced by the slowest neutrons. They are so slow that their speeds are comparable to those of gas molecules. The term **"thermal neutrons"** was therefore coined. Unlike the last nuclei mentioned, the nucleus $^{238}_{92}U$ requires much faster neutrons to start its fission. In short, this makes the process less controllable, with the potential for danger that has been avoided for the time being in nuclear power plants.

But let's get back to the cores actually used, $^{239}_{94}Pu$ e $^{233}_{92}U$. They can be produced with high efficiency, starting from $^{238}_{92}U$ and $^{232}_{90}Th$ respectively. These elements absorb neutrons and decay under emission β⁻ accordingly:

$$^{238}_{92}U \;+\; ^{1}_{0}n \;\xrightarrow{\;\beta^-\;}\; ^{239}_{94}Pu$$

$$^{232}_{90}Th \;+\; ^{1}_{0}n \;\xrightarrow{\;\beta^-\;}\; ^{233}_{92}U$$

The reactions are carried out in a "*breeder* **reactor**" (English: *fast breeder*; German: *Brutreaktor*) which will be discussed in detail on p. 118. The step of separating the plutonium from the unreacted uranium is known as the "Purex process"; separating the uranium from the unreacted thorium is known as the " Thorex process ".

In order to cause fission in nuclei that are lighter than the actinides, however, much more energetic projectiles are required. Thus, the fission of bismuth ($_{83}Bi$) and lead ($_{82}Pb$) can only be achieved with 100 MeV neutrons (lower limit: 50 MeV); fission of thallium ($_{81}Tl$) with 200 MeV deuterons, and the fission of platinum ($_{78}Pt$) and tantalum ($_{73}Tl$) only with 400 MeV particles. The more stable the target nucleus, the more energy the projectile must have. The fission products of these elements, unlike the $_{92}U$ (Figure 20) and $_{94}Pu$ have approximately the same atomic masses (= breakdown of the target nucleus in the medium). For the purpose of generating energy (p. 101) the fissions of these elements <u>are</u> useless. The considerable kinetic energy of the projectiles exceeds the "enthalpy" and "yield" factors of fission, which prevents a chain operation.

Spontaneous fission. The most obvious characteristic of actinides is the fission of their nuclei under neutron emission. This fission becomes easier and easier the heavier the element. Spontaneous fission occurs in the following isotopes:

$^{230}_{90}Th$, $^{232}_{90}Th$, $^{232}_{92}U$, $^{234}_{92}U$, $^{235}_{92}U$, $^{236}_{92}U$, $^{238}_{92}U$, $^{237}_{93}Np$, $^{235}_{94}Pu$ even $^{244}_{94}Pu$, $^{232}_{95}Am$, $^{234}_{95}Am$, $^{237}_{95}Am$ to $^{246}_{95}Am$, $^{240}_{96}Cm$ to $^{246}_{96}Cm$, $^{248}_{96}Cm$, $^{250}_{96}Cm$, $^{242}_{97}Bk$, $^{244}_{97}Bk$, $^{245}_{97}Bk$, $^{246}_{98}Cf$, $^{248}_{98}Cf$ even $^{250}_{98}Cf$, $^{252}_{98}Cf$, $^{254}_{98}Cf$, $^{253}_{99}Es$, $^{244}_{100}Fm$, $^{246}_{100}Fm$, $^{248}_{100}Fm$, $^{246}_{100}Fm$, $^{250}_{100}Fm$, $^{252}_{100}Fm$, $^{254}_{100}Fm$ even $^{258}_{100}Fm$, $^{252}_{102}No$, $^{254}_{102}No$, $^{256}_{102}No$, $^{258}_{102}No$, $^{259}_{102}No$.

Evidently, they are isotopes with even numbers of protons and neutrons that undergo fission particularly easily. These numbers then represent the opposite of the "magic numbers" mentioned on p. 52they make nuclei especially fragile. Here are some half-lives from this long list of isotopes:

$^{230}_{90}Th$	$> 1.5 \cdot 10^{17}$ years	$^{242}_{94}Pu$	$7.45 \cdot 10^{10}$ years
$^{232}_{92}U$	$3.5 \cdot 10^{13}$ years	$^{248}_{98}Cf$	$4.1 \cdot 10^{4}$ years

$^{236}_{92}U$	$2 \cdot 10^{16}$ years	$^{250}_{98}Cf$	$1.66 \cdot 10^4$ years
$^{238}_{94}Pu$	$2.6 \cdot 10^{10}$ years	$^{252}_{98}Cf$	85.5 years
$^{240}_{94}Pu$	$1{,}170 \cdot 10^{11}$ years	$^{254}_{100}Fm$	228 days

Nuclear fusion

After seeing how quite heavy nuclei undergo fission, as described in the last chapter, forming fragments of lighter nuclei in a very exothermic reaction, we have to realize that there is also the nuclear reaction of **fusion,** also quite exothermic, which unites lighter nuclei and transforms them into heavier ones. Both paths, fission and fusion, are exothermic, since the cohesive energy in the nucleus produced is higher. As Figure 14 the nuclei of medium-sized elements, i.e. those in the middle of the TPE, have maximum stability. Therefore, $_{26}$ Fe, $_{27}$ Co and $_{28}$ Ni can be considered the most stable elements when it comes to cohesive energy per nucleon; these elements are atypically frequent in the universe (the interior of the Earth is made up of them!).

The exotherm of uranium fission, as presented on p. 104 is unbeatable, in terms of kJ$\cdot$ mol^{-1} . However, if the reference is the weight of the reactants (kJ g^{-1}), nuclear fusion has an exothermia of an even greater order of magnitude than fission - since 1 g = 1 mol of hydrogen corresponds to 1 g = 1/235 mol of uranium.

Nuclear fusion, as we have already seen on p. 19is fundamental to the brilliance of the stars. The light of the Sun is closely linked to the exotherm of a cascade of reactions which, in the end, we have to classify as nuclear fusion. The starting element of the entire Universe is supposed to be hydrogen - at the same time the most abundant element in the Universe by far. We can therefore abstract the main fusion reaction as follows:

$$4\ {}^{1}_{1}H^{+}\ \rightarrow\ {}^{4}_{2}He^{2+}\ +\ 2\ {}^{0}_{+1}e^{+}\ +\ 2\ \nu_{e}\ +\ 26{,}7\ \text{MeV} \quad \text{Equation 30}$$

According to this summary equation, with each mole of He (= 4 g) produced, the star also develops 26.7 million faradayvolts = 2580 million kJ. This is about 8 times more than in the fission of the same mass (4 g) of uranium $^{235}_{92}U$ ($\frac{4}{235} \cdot 2 \cdot 10^8 = 3{,}404 \cdot 10^6$ FV = 329 million kJ).

In less hot stars ($\sim 10^7$ degrees) the mechanism can be described by a cycle known as the "proton-proton cycle" or "pp cycle" (*Ch. Critchfield*, 1938), shown in Equation 31. In hotter stars ($\sim 5 \cdot 10^7$ degrees) "catalysts" are still involved, namely carbon and nitrogen. The cycle is known as the "proton-carbon cycle" or "CNO cycle", shown in Equation 32 (*H. Bethe, C.F. v. Weizsäcker,* 1937). Our Sun belongs to the second category. The two cycles have already been abstracted in Figure 3 (p. 20).

Reagents		Products			Energy (MeV)	Time	Factor
^1_1H + ^1_1H	$\rightarrow$	^2_1H + $^0_{+1}e^+$ + ν_e		+	1,44	$14 \cdot 10^9$ years	2
^1_1H + ^2_1H	$\rightarrow$	^3_2He		+	5,49	6 seconds	2
^3_2He + ^3_2He	$\rightarrow$	^4_2He + $2\,^1_1H$		+	12,85	10^6 years	1
$4\,^1_1H$	$\rightarrow$	^4_2He + $2\,^0_{+1}e^+$ + $2\nu_e$		+	26,71	Equation 31	

^1_1H + $^{12}_6C$	$\rightarrow$	$^{13}_7N$	+	1,95	$1.3 \cdot 10^7$ years	
$^{13}_7N$	$\rightarrow$	$^{13}_6C$ + $^0_{+1}e^+$ + ν_e	+	2,22	7 min.	
^1_1H + $^{13}_6C$	$\rightarrow$	$^{14}_7N$	+	7,54	$2.7 \cdot 10^6$ years	
^1_1H + $^{14}_7N$	$\rightarrow$	$^{15}_8O$	+	7,35	$3.2 \cdot 10^8$ years	
$^{15}_8O$	$\rightarrow$	$^{15}_7N$ + $^0_{+1}e^+$ + ν_e	+	2,71	82 s	

$$^{1}_{1}H + \,^{15}_{7}N \rightarrow \,^{13}_{7}N + 4{,}96 \qquad 1.1 \cdot 10^5 \text{ years}$$

$$4\,^{1}_{1}H \rightarrow \,^{4}_{2}He + 2\,^{0}_{+1}e^{+} + 2\,\nu_e + 26{,}73 \qquad \text{Equation 32}$$

The times indicated in the partial reactions above should be interpreted as follows:

First line: the time that passes before a proton gets close enough to another proton so that they can react to form a deuterium, this takes on average $14 \cdot 10^9$ years.

Second line: it takes an average of 6 seconds for a deuterium nucleus to join with a hydrogen nucleus to form a helium nucleus.

Third line: after an average of one million years, two isotopes of $^{3}_{2}He$ are close enough to react to form $^{4}_{2}He$.

In the CNO cycle that leads to Equation 32, it can be seen that carbon and nitrogen are not consumed, but are actually reproduced. This is perfectly the definition of a **catalyst**. Otherwise, C and N inside the Sun would have been consumed long ago and the thermonuclear reaction would have stopped. However, the quantities of C and N inside the Sun must be quite small, because the average times in which the elementary reactions of the proton-carbon cycle take place are quite long. But it doesn't matter: the gigantic quantities of hydrogen available in the Sun compensate for this slowness, as we'll see in the calculation of the rate below.

The Sun not only produces positrons, but also neutrinos, ν_e , which are emitted at high speed into space[29] . They reach the Earth without much hindrance. It has been calculated that every square centimeter of the Earth is penetrated by $6 \cdot 10^{10}$ neutrinos from the Sun every second!

The radiation power of our Sun is, as anyone who has ever been burned knows, very high: each square meter of the Sun emits 61,300 kW = 61,300 kJ s^{-1} of

---

[29] This is the "opposite" of what happens in nuclear reactors here on earth, where reactions are accompanied by radiation from negatrons, β^- , as well as anti-neutrinos, $\overline{\nu}_e$.

energy. As there are $6.072 \cdot 10^{18}$ m^2 of surface area, this results in a total intensity of $3.72 \cdot 10^{23}$ kW $= 3.72 \cdot 10^{23}$ kJ s^{-1} .

This energy, as we have seen, comes from the fusion reaction, Equation 35. To provide all this energy, around 600 million tons of hydrogen ($5.8 \cdot 10^{14}$ g) must be consumed - every second! But don't worry: the Sun's hydrogen reserves are not small: around 10^{33} g. According to this estimate, the Sun has consumed only 1/50th of its hydrogen reserves every billion years since its creation ($\sim 6 \cdot 10^9$ years). It still has a long life ahead of it...

Another energy estimate for the Sun refers to its loss of mass, since the emission of radiation is equivalent to mass, according to the theory of relativity. In one billion years, the period in which it emits $1.2 \cdot 10^{40}$ kJ, it loses $1.3 \cdot 10^{23}$ g. This mass appears more figurative when expressed in tons per second: the Sun loses $4.1 \cdot 10^{12}$ g $= 4.1$ million tons every second! But this, after all, is only a small fraction (10^{-7} parts only) of its total mass ($\sim 2 \cdot 10^{27}$ t; see values in Table 7).

For now we're only talking about the synthesis of helium, which takes place at around 10^7 degrees, but what about the heavier elements? Well, the synthesis of the next elements requires even higher temperatures, in the order of 10^8 degrees. From there, the nuclei of ^4_2He nuclei can fuse into the next elements, for example:

$$^4_2He \quad + \quad ^4_2He \quad + \quad 0,095 \text{ MeV} \quad \rightarrow \quad ^8_4Be$$

$$^4_2He \quad + \quad ^8_4Be \quad \rightarrow \quad ^{12}_6C$$

In this way, elements of the type $(^4_2He)_n$ such as ^8_4Be, $^{12}_6C$, $^{16}_8O$, $^{20}_{10}Ne$, $^{24}_{12}Mg$, $^{28}_{14}Si$, $^{32}_{16}S$, $^{36}_{18}Ar$ e $^{40}_{20}Ca$. The elements located between these are the result of simple nuclear reactions of the $(^4_2He)_n$ as described in the chapter "Simple nuclear reactions". Elements heavier than these are formed at even higher temperatures: 10^9 to $5 \cdot 10^9$ degrees. In these extreme conditions, the nuclei of H, He, neutrons and γ rays assume energies and speeds comparable to the particles in a cyclotron or nuclear reactor. It is therefore not difficult to imagine that reactive collisions lead to new elements and isotopes of all kinds.

The synthesis of new elements here on Earth was completed around 5 billion years ago. But there is plenty of evidence that the synthesis of new elements occurs constantly in the Universe. For example, in some star systems the element $^{99}_{43}Tc$. As the half-life time is small, $t_{1/2} = 2.12 \cdot 10^5$ years, when compared to the age of the star system (10^7 years), it is clear that this element is continually being produced.

Interestingly, in certain types of "Novae" (p. 139) the emission of the first light is very strong, but then, with a half-life of ~55 days, it diminishes. This led astronomers to search for isotopes with this $t_{1/2}$. In fact, this $t_{1/2}$ is quite rare, and is made by the element $^{254}_{98}Cf$ ($t_{1/2} = 60.5$ days) which decays spontaneously, but also by the isotope ^7_4Be which is transformed into $t_{1/2} = 53.4$ days under K capture at ^7_3Li.

There are still many doubts and facts about the genesis of stars that have not (yet) been conclusively explained. But today there is unanimity that the birth of stars begins with an occasional accumulation of interstellar matter (hydrogen). This matter then densifies in the center due to gravitation. This accumulation is autocatalytic: the heavier the center, the greater the force with which it attracts new matter from space. As there is no exchange of energy with the environment, this densification is accompanied by adiabatic heating. The heating continues and finally reaches 10^7 degrees, creating favorable conditions for the synthesis of He according to Equation 32. This synthesis is exothermic, so it creates favorable conditions for the formation of the other elements.

The end of a star is also studied. When all the hydrogen inside the star has run out, the helium formation moves outwards in waves towards the star's surface. A pulsation is observed and then the star expands, becoming a "red giant". At this stage the mass in its center accumulates and the temperature rises. Conditions gradually reach (> 10^8 degrees) where medium-mass elements are formed from the fusion of helium nuclei, as described above. At the end of this period, the star gradually contracts and becomes a "white dwarf ". Inside the star, temperatures can reach $5 \cdot 10^9$ degrees, creating the conditions for the formation of the heaviest elements, syntheses that are endothermic in nature. After this phase, the white dwarf loses temperature and cools down over a few million years.

A special event can disrupt the final stage of a star. If the synthesis of elements occurs in such a way that the thermal balance is disturbed (i.e. more exothermic synthesis occurs than endothermic synthesis), the star can overheat and explode. This event is called a "nova" or "supernova" (p. 141). And there is an even rarer - and more catastrophic - event. This is observed only in especially large stars, >150% the size of our Sun. These stars, when they contract, can create gravitational fields so strong that they go beyond the "white dwarf" state. Under these unimaginable pressures, the protons and electrons in the star's plasma fuse to form neutrons. This is the beginning of a neutron star, also called a "pulsar".

In the worst case, these pulsars create gravitational fields so high that total collapse occurs. The gravitation is transformed into energy and a "black hole" is formed.

Humanity will benefit greatly from nuclear fusion, as it happens inside the Sun. If it were possible to achieve the fusion of H forming He here under well-controlled conditions, the problems of energy here on Earth would be solved once and for all. Research into nuclear fusion has been going on for around 50 years; for the time being, the energy expended to create favorable and controllable conditions is greater than the energy that results from the exothermia of fusion. Even so, technological advances are expected in this area, see p. 126.

Unfortunately, humanity has already achieved uncontrolled nuclear fusion: the hydrogen bomb, which is still 10 times more devastating than the "ordinary" nuclear bomb (p. 144). 141).

The nuclear chain reaction

On p. 101 the nuclear fission of $^{235}_{92}U$ which leads to exactly two new nuclei of unequal mass, 2 to 3 neutrons and a lot of energy. While on p. 103 the focus was on the fragments of the nucleus and the energy of the nuclear reaction, here we will focus on the neutrons produced. It is precisely the neutrons released in this reaction that can lead to the continuation of this reaction in other nuclei $^{235}_{92}U$ A kinetic chain reaction. In order to take advantage of this chain reaction, in the form of nuclear power plants, a series of conditions must be met, as will be explained in detail below. If you "harness" > 1 reactive neutron per elementary act, then you lose control because an avalanche of neutrons is produced, the reactor quickly goes out of control and explodes violently.

Before going into detail, however, we should honor the immense amount of research and experimentation that this nuclear reaction cost before it became useful to humanity. It was in the United States, at the time of the Second World War. The Hollywood movie *"Oppenheimer"* (2023) sketches the efforts made at the time. In just a few years, a lot of complex problems were solved that, under normal conditions, would have taken an entire generation of researchers and engineers. Soon after *O. Hahn*'s discovery of nuclear fission, Germany launched a war. It wasn't long before physicists around the world recognized the military potential of this violent reaction. At the same time, there was great fear that the country where the reaction was discovered would take advantage of it in the form of new weapons. That's why the US released funds for research and technological development like never before. At an impressive rate, more than 10,000 researchers[30] and engineers worked together on the " *Manhattan Project*". The military director of the project was General *Leslie R. Groves,* and the scientific director of the research work was physicist *J. Robert Oppenheimer.* More than 150,000 people worked directly or indirectly on the project in the utmost secrecy. The investment amounted to several billion dollars and culminated, on December 2, 1942, in the first working nuclear reactor in Chicago. The first time that a reactor was fully sustained by the chain reaction was called the "critical moment of the reactor". This was followed in 1945 by the explosion of the first atomic bomb in the state of New Mexico, USA, and shortly afterwards its application in two Japanese cities. British and Canadian scientists, who were developing their own nuclear weapons project under the code name *"Tube Alloys"*, cooperated with US researchers from 1943

---

[30] The most important names are listed on https://de.wikipedia.org/wiki/Kategorie:Person_ (Manhattan-Projekt).

onwards ("Quebec Agreement"). By the end of 1945, the cost of nuclear research and development in the US alone was estimated at $1.9 billion.

Control of nuclear chain fission

Natural uranium is a mixture of a lot of the isotope 238 (99.3%), very little of the isotope 235 (0.7%) and almost none of the isotope 234 (0.006%). When you attack this natural mixture with neutrons, there will be no chain reaction - although each elementary act will produce 2 to 3 neutrons, depending on the act:

$$^{235}_{92}U + {}^{1}_{0}n \rightarrow {}^{236}_{92}U \rightarrow X + Y + 3\,{}^{1}_{0}n \qquad \text{Equation 33}$$

X and Y being the two fragments of unequal mass, as described on p. 103. Why doesn't the reactivity spread? There are two answers:

1) Because the neutrons formed don't have the right energy;
2) Because they are consumed in another reaction that doesn't propagate in a chain.

To understand these answers, we have to recognize that the two isotopes of uranium, $^{238}_{92}U$ e $^{235}_{92}U$ have completely different nuclear behaviors. Only $^{238}_{92}U$ takes advantage of the neutrons in Equation 33. Through the reaction

$$^{238}_{92}U + {}^{1}_{0}n \rightarrow {}^{239}_{92}U \; (\xrightarrow{\beta^-} \xrightarrow{\beta^-} {}^{239}_{94}Pu) \qquad \text{Equation 34}$$

and the abundance of the isotope 238, the neutrons are completely consumed, i.e. unavailable to sustain the reactions of Equation 33. For every slow neutron used as a projectile, only one elementary act of reaction 33 takes place. That's good, because we still have uranium on Earth. If a chain reaction were possible, then neutrons from cosmic resources would have caused the atomic explosion of natural uranium a long time ago.

The other aspect is the energy of the neutrons. The absorption reaction, Equation 34, occurs especially easily with 25 eV neutrons (this corresponds to a speed of > 70 Km s^{-1}). On the other hand, the reaction 33 that leads to fission requires neutrons of energies 1000 times lower: only 0.0025 eV, corresponding to a speed of 2.2 Km s^{-1} (this is approximately the speed of gas molecules at room temperature).

The reason why isotopes $^{238}_{92}U$ e $^{235}_{92}U$ isotopes behave so differently is to be found in the structure and stability of these nuclei. O $^{235}_{92}U$ obviously has an odd number of neutrons, the $^{238}_{92}U$ an even number. We know that neutrons like to group themselves in pairs (the same goes for protons). The nucleus of the isotope 238 is therefore more accommodating, more stable. For this reason, the union of $^{235}_{92}U$ with a neutron releases more energy (6.8 MeV) than the absorption of a neutron in the $^{238}_{92}U$ (5.5 MeV).

The nuclear fission stage, the right-hand part of Equation 33, requires an activation energy of of 6.5 MeV. To cause fission of the nucleus $^{239}_{92}U$ would require an activation energy of 7.0 MeV. Now you can see that in the case of absorption

$$^{235}_{92}U \;+\; ^{1}_{0}n \;\rightarrow\; ^{236}_{92}U$$

Slow neutrons are adequate, because the activation of the subsequent fission is guaranteed. On the other hand, slow neutrons would not have enough energy to induce the fission of^{239} U. In this case, you really need fast neutrons whose kinetic energy will complete the missing value to trigger fission: 7.0 - 5.5 = 1.5 MeV.

Similar considerations apply to explain the <u>easy</u> fission of:

$^{233}_{92}U$ (Absorption energy: 7.0 MeV; fission activation energy: 6.0 MeV) and

$^{239}_{94}Pu$ (Absorption energy: 6.6 MeV; fission activation energy: 5.0 MeV)

and also the <u>difficulty</u> in fissioning thorium:

$^{232}_{90}Th$ (Absorption energy: 5.4 MeV; fission activation energy: 7.5 MeV).

In order to suppress the parallel reaction 34, substances that slow down the neutrons produced are used. The so-called "**moderators**" should reduce the speed of recent neutrons (~2 MeV, corresponding to 20,000 Km s^{-1}), below the critical value of 25 eV, as quickly as possible. Only then are they no longer being consumed by the parallel reaction 34. There are different materials that decelerate fast neutrons.

The most widely used and approved moderators are heavy water, D_2 O, and carbon in the form of graphite. For a moderated nuclear power plant with heavy water, 50 - 80 tons of D_2 O are calculated for every 100 megawatts of power. For the plant, heavy water is just an initial investment, as the D_2 O is driven in a closed circle. A much cheaper alternative would be ordinary water ("light water plant"). Graphite is not as effective as a moderator, but much cheaper than heavy water. However, this graphite must be of good quality, i.e. free of impurities that absorb neutrons. After all, neutrons should only be slowed down by elastic shocks with the moderators, but not absorbed. The technical solution could be a graphite bed in which the purified uranium rods (= UO_2 = nuclear fuel) are arranged in parallel. The situation is sketched in Figure 21.

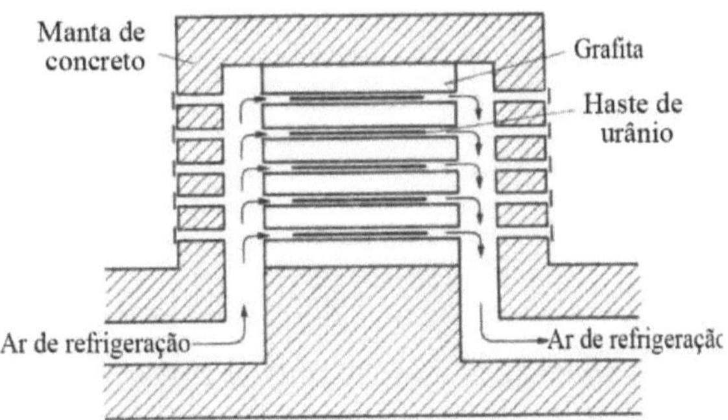

Figure 21 Schematic of a uranium nuclear reactor (there are many other variations).

It is important to know that there is a minimum size for the nuclear reactor, the " **size critical**". This is due to the [surface : volume] ratio, which decreases as the size increases. A smaller surface area means that fewer neutrons escape outwards, i.e. more neutrons collide with the fuel material. They also use "reflectors ", made of heavy water, graphite or beryllium, to reduce the rate of neutrons escaping from the package. They manage to some extent to divert the neutrons so that they stay inside the reactor. Only a reactor above the critical volume has enough neutrons (a portion of which are actually absorbed by the $^{238}_{92}U$) to keep the chain kinetics of reaction 33 running.

The nuclear event is abstracted in Figure 22. Of the three neutrons formed in elementary act 33, all are stopped below the critical speed, then two are eliminated, either by absorption according to reaction 34, or by escaping out of the block. Only one neutron, better to say <u>exactly</u> one neutron, is used to give continuity, by being absorbed by the $^{235}_{92}U$ according to Equation 33.

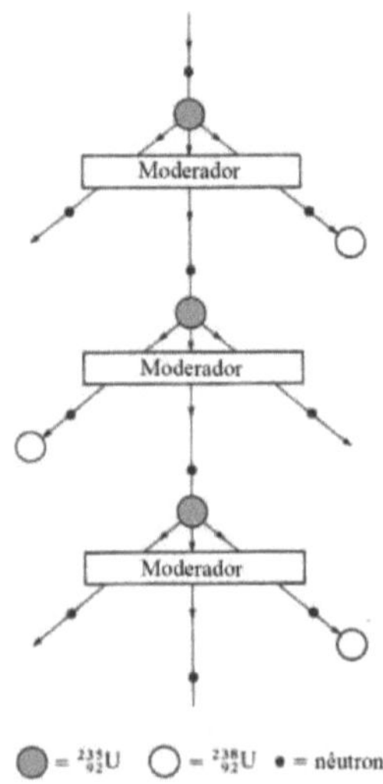

$\bullet = ^{235}_{92}U$ $\bigcirc = ^{238}_{92}U$ $\bullet$ = néutron

Figure 22 Schematic of the controlled chain reaction.

The number of neutrons released in the elementary act is only an average value. In addition, many factors influence the speed of the chain reaction. Therefore, every nuclear reactor needs a device to finely regulate the rate of incoming neutrons, the **"moderators"**. A common method are movable rods, made of boron or cadmium steel, which can be pushed between the fuel rods or pulled out. In the first case, the reaction rate is lowered, in the second case it is increased, as these materials are excellent neutron absorbers. Incidentally, this is also why uranium used as fuel must be especially pure, i.e. free of neutron-absorbing elements (physicists speak of impurities with large absorption radii for neutrons). Finally, these devices can be used to regulate the temperature of the reactor, according to the demand for energy used to generate steam for the turbine generators.

In a typical nuclear reactor, there are 40 - 50 tons of uranium rods densely packed, wrapped in 200 - 400 tons of graphite. There are around 1000 uranium rods, each 2 - 3 cm in diameter and 6 m long, with a uniform distance of ~20 cm. Sequestering the immense amount of heat and keeping the reactor sufficiently cool are not easy tasks. A nuclear power plant can develop up to 1000 MW of power, i.e. its exotherm is one million kJ s^{-1} . The rods are therefore placed in coaxial channels through which the cooling medium also passes. What was said about the moderator material and the fuel material also applies to the cooling medium: it must not absorb neutrons. Most power plants use water or high-flow air.

A " multiplication factor " k is useful for describing the conditions under which a nuclear power plant can operate safely. The factor is simply the ratio of neutrons produced to neutrons consumed in reaction 33. In the value of neutrons produced, only those that actually remain inside the reactor count, those that escape the reach of the rods and those that have been absorbed by the reactor do not count. $^{238}_{92}U$ do not count. The neutrons consumed include only those that induce a new reaction 33, so those that have the right energies and directions to fuse with a nucleus. $^{235}_{92}U$. Safe operation therefore requires this multiplication factor to be strictly 1. If $k < 1$, the kinetic chain dies and the reactor cools down. Worse would be a factor of $k > 1$, because in this situation you would have to count on an exponential growth of reactive neutrons in the reactor, an avalanche and explosion would be the consequence. This is what happened, for different reasons, in the Chernobyl (Ukraine) and Fukushima (Japan) disasters. Control, as mentioned above, is achieved by means of neutron absorbers. The adjustment of their position is automated and works via an ionization chamber where the density of reactive neutrons is analysed. The fact that a small part of the fission neutrons is released with a delay makes it much easier to control and adjust the absorbers. Thus, these devices do not need to be adjusted in fractions of a second, but within a few minutes, without the reactor running out of control.

Today's nuclear reactor represents an important generator, not only of energy that involves few fossil resources, but it is also a factory that provides new (radioactive) material for various technological areas. The generation of $2 \cdot 10^8$ faradayvolts = $1.9 \cdot 10^{10}$ kJ$\cdot$ mol^{-1} = $8.1 \cdot 10^{10}$ kJ per Kg of uranium is used to continuously produce steam, driving turbines to generate electricity. In addition to this thermal energy, let's not forget the kinetic energy of the neutrons emitted from the reactor, which are thus the most intense of all neutron sources (p. 73). The generation of neutrons in this reactor can be increased to more than 10^{15} neutrons$\cdot$ cm s$^{-2,-1}$. It can thus produce valuable new isotopes of all the TPE elements, on a gram scale. There are two possibilities: either the elements to be transformed are placed in the vicinity of the clump of rods, where they benefit from the neutrons escaping from the reactor. Or they are introduced into the reactor via a probe. Of course, the absorption of neutrons by these elements must be compensated for, so the absorber rods must be pulled out a little.

And let's not forget the fission products either. The controlled fragmentation of uranium provides the elements lanthanides, actinides (including Pu), as well as Kr, Sr, Zr, Mo, Tc, Ru, I and Sc, on a kilogram scale. All of them are radioactive.

In this context, it's worth remembering the mass-energy equivalent, as shown in Equation 10 (p. 64). However, although the energy of fission is enormous, the loss in mass is not very significant: from 1 kg of uranium, 999 g of fragmented elements are obtained.

Breeder reactors

The generation of plutonium in the uranium reactor deserves another word, because this artificial element is of great importance as a source of alternative energy: it also serves as nuclear fuel. The $^{239}_{94}Pu$ from natural uranium (possibly enriched in $^{235}_{92}U$) is known as a **plutonium** breeder **reactor** (alternative name: breeder reactor). The neutrons from the fission of $^{235}_{92}U$ when they have the appropriate energy (25 eV), can be absorbed by the nucleus $^{238}_{92}U$ making reaction 39. O $^{239}_{92}U$ is radioactive and decays in $t_{1/2}$ = 23.54 minutes under radiation β⁻ forming neptunium. This happens because the ratio [neutrons : protons] is much higher than the ideal value, so in radiation β⁻ a neutron is transformed into a proton and the ratio improves. O $^{239}_{93}Np$ is also unstable ($t_{1/2}$ = 2,355 days) and finally decays into plutonium, $^{239}_{94}Pu$ which is less unstable than the intermediate nuclei ($t_{1/2}$ = 24,390 years). This makes it much easier to handle. The whole series (= Equation 34 written out in full):

$$^{238}_{92}U \xrightarrow{+n} {}^{239}_{92}U \xrightarrow{-\beta^-} {}^{239}_{93}Np \xrightarrow{-\beta^-} {}^{239}_{94}Pu \qquad \text{Equation 35}$$

A fixed graphite sleeve (moderator) around the uranium rod, as is the case in the ordinary uranium reactor, would not be useful in the case of the breeder reactor. This construction will prevent the continuous operation of reaction 35. In the case of plutonium generation, uranium is used in the form of short, thick rods (length 15 cm, diameter 5 cm) which are placed in aluminum tubes. The tubes are well cooled by water or air. The rods are moved through the moderator. This makes it possible to load the tube with combustible material on one side, and unload the products containing the plutonium on the other side. In one run, Pu production is around 0.1%.

The separation of *plutonium* from unreacted *uranium* takes place through the *Purex* process (*Plutonium-Uranium-Extraction*), which was already mentioned on p. 105. The mixture is first dissolved and oxidized in strong nitric acid. This solution is then extracted by 30% tributylphosphate, where both U and Pu make complexes and go into the organic phase. The organic phase is treated with reductant and the Pu quickly changes to NOX +3 in which it becomes insoluble and can be re-extracted into the aqueous phase. During this stage, the U is reduced more slowly and can be re-extracted into the aqueous phase later.

These steps can only be carried out under maximum safety measures, since the nuclear reactor and the reaction products are very radioactive, and any direct contact would lead to instant death. This danger becomes even more apparent when we remember that the radiation β and γ, that is released from the products formed in the fission of 1 kg of U or Pu is equivalent to the radiation from 1000 tons of radium! That's why the nuclear reactor is surrounded by a thick concrete wall, impenetrable to gases. All the chemical steps, such as dissolution, oxidation, extraction, reduction, etc., are carried out in thick concrete cells, with remotely operated movable arms, while countless sensors and cameras monitor what is happening inside this "hot chemistry laboratory".

In short, a 1000 MW reactor can produce 1 kg of plutonium per day.

Another type of **breeder reactor** uses **thorium** instead of uranium. The neutrons needed for this process also come from the fission of $^{235}_{92}U$ or $^{239}_{94}Pu$. As seen above, the decay series runs through short-lived intermediate nuclei and ends in the relatively stable $^{233}_{92}U$ relatively stable ($t_{1/2} = 1.59 \cdot 10^5$ years):

$$^{232}_{90}Th \xrightarrow{+n} {}^{233}_{90}Th \xrightarrow{-\beta^-} {}^{233}_{91}Pa \xrightarrow{-\beta^-} {}^{233}_{92}U \qquad \text{Equation 36}$$

O $^{233}_{92}U$ today is the third most important nuclear fuel. But it is predicted that this process will become even more important in the future, as the starting material thorium is four times more abundant in the earth's crust than uranium. It has a further ethical advantage: this material is not used to produce the bomb. This applies to both the decay products and the nuclei created from thorium . The explanation lies in the difficult fission of thorium (p. 114)

The separation of $^{233}_{92}U$ from the reaction mixture in this breeder reactor is done using the *Thorex* (*Thorium Extraction*) process, which works in a similar way to the *Purex* process described above. After dissolving the mixture in strong nitric acid, extraction follows, this time with 40% tributylphosphate. In this, only Th and U go into the organic phase. From this phase, when treated with dilute nitric acid, the metals are extracted back, first Th, then U. Finally, the metals can be isolated from the aqueous solutions, either by precipitation or ion exchange.

The interesting thing about breeder reactors, both thorium and plutonium, is that they produce more fissionable material than they consume. Simplified, this is due to the production of 2 - 3 new neutrons with each elementary act, while only one of them will be used to continue the chain reaction. The rest (apart from those that escape from the reactor without reacting) can be used to induce transformations, $^{232}_{90}Th \rightarrow {}^{233}_{92}U$ e $^{238}_{92}U \rightarrow {}^{239}_{94}Pu$ respectively. For each fissioned nucleus, > 1 new fissile nucleus can be produced. It is common to characterize the power of a breeder reactor by its "doubling time", which describes the time it takes to have doubled the amount of fissionable material that has been invested in fuel material. This time is around 10 years.

There is also a third category of breeder reactors. New fuel material is produced at an extraordinarily high rate in the so-called **"breeder reactor fast "**. Fast neutrons convert the isotope uranium-238, which is unsuitable as nuclear fuel, into the new fissile fuel plutonium-239. This technology is (still) relatively rare, but could become more important worldwide if nuclear energy is expanded. Its operation and characteristics are outlined below.

Most current reactors for electricity generation use the isotope of^{235} U, which can only be split with neutrons that have been decelerated to thermal speeds (p. 115). Given that only 0.7% of^{235} U occurs in nature, most of the uranium remains unused for energy production. This is because^{238} U cannot be used as a fissile material in thermal neutron reactors.

But^{238} U has another property that can be used to generate energy: it can capture a fast neutron, such as the one released immediately after a fission process. It can then be converted into^{239} Pu through various decay processes, described in Equation 35. This isotope, in turn, is a good fission material, even for fast neutrons. One disadvantage: plutonium can also be used to build nuclear weapons.

Thus, two important processes take place at the same time: on the one hand, energy is generated through the fission of^{239} Pu and, on the other hand, the new fission material^{239} Pu is generated from non-fissile uranium,^{238} U. Since this type of breeder reactor works with fast neutrons and produces new fissile material, it is called a "fast breeder reactor".

The core of such reactors can be designed in such a way that more^{239} Pu is produced from^{238} U than is consumed by plutonium fissions. In this way, a surplus of fissile material is obtained, which can be used to generate energy in other reactors. By *upgrading* this process, it is possible to use natural uranium around sixty times better than light water reactors (which only use^{235} U). In other words: the energy potential of uranium can be increased by a factor of 60 using this technology.

As breeder reactors use fast neutrons to generate energy, these particles must not be slowed down after the plutonium has been split. The moderators that are so useful in "thermal" reactors are no longer an option for dissipating the heat released. A technical solution offers liquid sodium that can be used as a coolant, as in the former French *Superhénix* breeder reactor (duration: January 15, 1986 to December 31, 1998) or the Russian *Beloyarsk 4* breeder reactor (start-up: December 10, 2015). Metallic sodium has a melting temperature of 98 °C; its boiling temperature is 883 °C. As the temperatures at which sodium is used in fast breeder reactors are between 400 and 550 °C, the coolant is far from its boiling point, so these types of reactors operate under a relatively low pressure of only around 10 bar in the primary circuit. By comparison: water-moderated reactors have pressures of 160 bar. However, as you can imagine, using liquid sodium that is so reactive against oxidizers calls for even higher safety measures than in a conventional nuclear reactor.

According to the World Nuclear Association of July 2017, a total of eleven breeder reactors have been built in nine countries since 1962. By the end of 2016, eight of these plants had been closed, three are still in operation: in Russia the *Beloyarsk 3* reactors (560 MW, since April 1980), *Beloyarsk 4* (789 MW, since December 2015) and in China the *Cefr breeder reactor* (20 MW, since July 2011). At the end of 2016, the twelfth *"Pfbr" breeder* reactor with a capacity of 470 MW was under construction in India.

The generation of nuclear energy and new isotopes, as is the standard today, was only possible with the "good will" of nature. The 235 and 238 isotopes of uranium have half-lives ($0.7 \cdot 10^9$ and $4.5 \cdot 10^9$ years, respectively) that are of the same order of magnitude as the age of the Earth ($6 \cdot 10^9$ years). Thanks to these long half-lives, we still have these isotopes today (compare Figure 13) - no longer in equal quantities, but still in the ratio [0.7 : 99.3]. Just imagine that the $t_{1/2}$ of ^{235}U was only a third of the real time, so its share in the natural uranium mixture would today be only a millionth, i.e. the $^{235}_{92}U$ would only exist in traces. The only isotope would be $^{238}_{92}U$ which, in turn, is not fissile in a chain reaction and is therefore useless for generating energy and making new elements. After all, that was the fate of the transuranic elements: they all have remarkably short half-lives, so they practically disappeared from the Earth. Certainly, they were produced at the beginning of the planet in quantities comparable to U. But they all decayed more quickly and disappeared completely from the Earth. The only way to bring them back is by synthesis, as described above.

Nuclear reactor fuel material

"MOX " is the name given to mixed oxide fuel elements. These are fuel elements which, unlike fuel elements made from pure uranium dioxide, contain another oxide. This is usually plutonium dioxide, more rarely thorium dioxide.

Plutonium can be used in the form of MOX fuel in nuclear reactors and used there to generate energy. This is an option for light water reactors, but a necessity for fast breeder reactors (p. 120), because the high neutron yield required for nuclear fission is only achieved with a sufficiently high proportion of ^{239}Pu in the fuel (and fission by fast rather than moderate neutrons). Fast breeder reactors can also be operated with highly enriched uranium, in which the proportion of ^{235}U is sometimes 80-90%, instead of ~3-5% (as in light water reactors) or 0.72% (as in natural uranium reactors). Due to the high costs of preparing highly enriched uranium, however, this is less economical than using plutonium.

In the thermal neutron spectrum, a reproduction number > 1 is only possible when thorium is used. However, as ^{232}Th alone cannot sustain a chain reaction, it is necessary to use mixed fuel with an appropriate content of ^{235}U, ^{239}Pu or ^{233}U, or to use an external neutron source to operate as a subcritical reactor. ^{235}U can be obtained from natural uranium, while ^{239}Pu (from ^{238}U) or ^{233}U (from ^{232}Th) must be "created" (p. 118) and therefore extracted from the spent fuel of other reactors.

Separation of uranium isotopes

Separating the two isotopes is not an easy task - since they have identical chemical behaviors. So the method must be physical, taking advantage of the different masses, densities, velocities, diffusion coefficients, etc. And even these sophisticated methods have to deal with relative differences in the atomic masses of ^{235}U and ^{238}U of just 1.3%. Since 1942, various methods have been developed, among which fractional diffusion of a volatile uranium compound is the most widely used. You heard that right: the compound UF_6 has a vapor pressure of ~10 kPa at room temperature (sublimation at 56 °C); it's easier to imagine a bumblebee flying than this big guy, isn't it?

The term "enrichment" refers to the various processes used to increase the proportion of the isotope ^{235}U in uranium. Natural uranium consists of approximately 99.27% ^{238}U and 0.72% ^{235}U, as well as traces (55 ppm) of ^{234}U. Uranium enriched to various levels of ^{235}U is used as nuclear fuel for nuclear reactors, but also for nuclear weapons. Natural uranium is not suitable for use in ordinary light water reactors, but can be used in heavy water reactors and graphite-moderated reactors. Enrichment is therefore an important branch of the uranium industry, as it makes the operating modes of nuclear power plants more flexible.

Common industrial processes use uranium hexafluoride (UF_6) as the process medium, the only chemical compound in uranium that has sufficient volatility for the separation process at room or low temperatures. First, *yellowcake*, a mixture of various uranium compounds, predominantly oxides, is extracted from the uranium ore by leaching. Uranium hexafluoride is the product of converting the yellowcake with hydrofluoric acid.

The second argument in favor of uranium hexafluoride for the enrichment process: fluorine occurs in nature as a pure element (p. 51), the only isotope of which is ^{19}F. Thus, the mass of the UF_6 molecules varies only due to the different masses of the uranium isotopes. Although the atomic mass of fluorine is small, the difference in relative masses between UF^{235}_6 and UF^{238}_6 decreases at

$$1 - \frac{6 \cdot 19 + 235}{6 \cdot 19 + 238} \approx 0{,}0085 = 0{,}85\%$$

Gas centrifuge. Centrifugation has overtaken gas diffusion in importance for enriching uranium. It is the method that uses the least energy and offers the greatest flexibility in plant capacity.

The gaseous uranium hexafluoride (UF_6) is fed into a vertical cylinder that rotates at a very high speed (over 60,000 rpm). Under the influence of the high speed and the resulting mass-dependent centrifugal force, the heavier molecules of UF^{238}_6 accumulate on the inner wall of the cylindrical rotor and the lighter molecules of UF^{235}_6 a little closer to the rotor axis, so that the isotopes can be removed separately using *Pitot* tubes. The centrifugal force creates a pressure gradient. At maximum speed there is practically a vacuum on the shaft, so no sealing is required for the feed and extraction tubes.

The rotor itself operates at reduced pressure to avoid friction and is driven by a rotating magnetic field.

The separation effect is enhanced in modern centrifuges by stimulating an axial circulating flow. This can be generated by thermal convection, but today it is mechanically excited by one of the *Pitot* tubes. These centrifuges are called countercurrent centrifuges. The largest separation factor between the enriched and depleted mass flow with[235] U is no longer axial, but between the ends of the centrifuge. The enriched light fraction ("product") and the heavier fraction ("residue") are removed from these two ends of the centrifuge. The axial separation factor depends exponentially on the square of the peripheral velocity.

The separation process takes place under reduced pressure. Therefore, the product and residue must be brought to the appropriate pressure and temperature, which happens in sublimators / resublimators, before they can be placed in transport or storage containers.

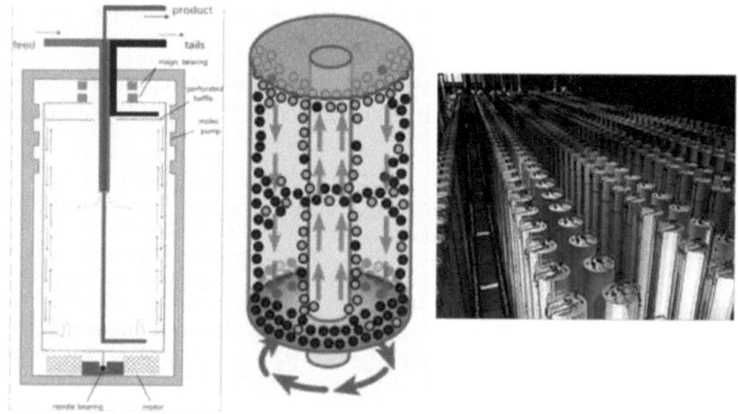

Figure 23   Gas centrifuges for separating uranium isotopes:
Left: *Zippe*   centrifuge.
Middle: counter-current gas centrifuge ($^{235}UF_6$ in light blue, UF^{238}_6 in dark blue).
Right: cascade of centrifuges in a medium-capacity plant.

Source: https://de.wikipedia.org/wiki/Uran-Anreicherung

Gas centrifuges are usually cascaded with several hundred individual centrifuges, since each centrifuge can only achieve a limited yield and enrichment. Connecting the centrifuges in parallel leads to an increase in throughput, while enrichment is increased by connecting them in series. The separation factor of a centrifuge can be increased by increasing the length of the tube and especially the revolutions (rpm). The usual diameters of the cylinders are 10-20 cm, the length is usually 50-100 cm; the speed at the inner wall reaches up to 750 m s^{-1}. For the TC-21 centrifuge from Urenco (USA), diameter 20 cm, length 5 m, peripheral speed 750 m/s, ratio [vertical circulation : feed] of 4, a separation factor [^{235}U :^{238}U] of 1.74 was calculated.

Gas diffusion. The diffusion process is older than the gas centrifuge, but it consumes a lot of energy: up to 60 times more! At the time of the *Manhattan Project* (p. 112), most of the US electricity generation was needed just to operate these plants.

Gaseous uranium hexafluoride can diffuse through a microporous membrane, with the pore diameters being smaller than the length of the free path in the gas (*Knudsen* flow). The driving force is the pressure difference on both sides of the membrane. Molecules containing the lighter isotope^{235}U have a higher diffusion rate (proportional to $M^{-1/2}$, where M is the molecular mass) than heavier ones, so the light isotopes accumulate in the low-pressure chamber. The gas that diffuses through the pores ("product") therefore contains a slightly higher proportion of the isotope^{235}U than the original flow ("feed"). A single stage has a low separation factor, a maximum of 1.0035, but a high material yield. To achieve an enrichment level sufficient to operate light water reactors (p. 122) you need around 1,200 stages connected in series.

Calutron. Using the principle of the mass spectrometer (electromagnetic separation), the isotopes of uranium can be clearly separated. First the uranium atoms are ionized, then accelerated uniformly in an electric field and then deflected by a magnetic field. The less heavy ion is deflected more. This method was used during the Second World War to produce enriched uranium for the first atomic bombs. However, due to the enormous effort involved, this process is no longer relevant to the production of enriched uranium today. However, it is used in research for other isotope separations because, ideally, only a single atom of an isotope can be detected (useful for characterizing the products of singular nuclear reactions; p. 70).

Use of enriched uranium. ^{235}U - like some other nuclides with an odd number of neutrons - is easily fissile by thermal neutrons and is the only naturally occurring nuclide capable of sustaining a chain reaction. Although natural uranium can also be used for heavy water reactors (D_2O) and graphite-moderated reactors, light water reactors (H_2O; p. 122) must be fed with uranium whose[235] U content has been increased to at least 3% - in practice, up to 5%. Uranium with 20% or more[235] U is called highly enriched uranium. Nuclear weapons require very high enrichment (usually at least 85%).

Use of depleted uranium. Depleted uranium is the by-product of enrichment. For every ton of nuclear fuel enriched for civilian purposes, around 5.5 tons of depleted uranium are produced, the[235] U content of which is still ~0.3%. Due to its high density, it is used as a weight to balance the wings of aircraft and racing yachts, as well as in military uranium ammunition. However, so far, only about 5% of the depleted uranium produced has been used for such purposes; the rest is stored. Since the isotope ratio is irrelevant for non-nuclear applications, depleted uranium is mainly used in other applications (chemical laboratory, tracers, among others). The main interest in this material, particularly in Russia, is its use as a blending material to convert highly enriched uranium (military) into low enriched uranium (civilian); only after this conversion is the mixture used in light water reactors. Special mention should be made here of the disarmament campaign under the START II Agreement: "*Megatons to Megawatts*". It can also be used to produce fuel elements in breeder reactors (p. 118) and as "MOX" (p. 122).

Towards controlled nuclear fusion

Unlike fission, it has not yet been possible to use nuclear fusion for civilian purposes. Accelerating and focusing hydrogen costs more energy than fusing it to helium. But let's not lose hope, because the day this technology is mastered, humanity's energy problems will be solved forever. Since the nuclear fuel for this reaction, hydrogen, is available in unlimited quantities in the form of water. While coal, oil, gas and uranium are limited fossil resources, seawater really won't run out.

In space, nuclear fusion takes place in a controlled manner inside stars, including the Sun. For billions of years, it has provided thermonuclear energy in a fairly reliable and constant way (p. 106). In the uncontrolled course of both nuclear fission and fusion, the reaction gets out of hand and leads to a catastrophe, an explosion of gigantic or even cosmic size. On Earth, this has already happened (the atomic bomb and the hydrogen bomb, respectively); in the Universe, it occurs in "nova" or "supernova" events. More on these cases in the last chapter...

Figure 24   Reduced pressure chamber of the JET fusion reactor, Culham, Great Britain.

Source: https://www.tagesschau.de/wissen/technologie/energierekord-jet-100.html

Replicating the energy source of stars on Earth - that's the goal of research into nuclear fusion reactors. It would be our ideal energy source for the following reasons:

- Cheap combustible material in unlimited quantities, as mentioned above;
- Small amounts of waste that are weakly radioactive;
- There are no chain kinetics, because fusion produces at most one fast neutron. This has the advantage that an explosion, as in the case of nuclear fission, does not occur. But it also has the disadvantage of being easily extinguished if the conditions inside the reactor get out of control.
- There are no long-lived radioactive products (which are produced by fission in large quantities) that cause environmental problems.

Atomic basis of fusion

The most promising concepts for fusion reactors to date involve a deuterium-tritium plasma, in the [1 : 1] mixture, in a ring-shaped magnetic field and heating it to a very high temperature. To achieve a net energy gain in this way, the plasma volume must be large enough.

Instead of neutrons (thermal or fast, depending on the type of fission reactor), it is an extremely high temperature that ensures the continuation of nuclear fusion. The only reaction, within a series of reactions between the isotopes of hydrogen

(Equations 38), that can be used for the time being is the fusion of deuterium (D $=^2$ H) with tritium (T $=^3$ H), as follows

$$D + T \rightarrow ^4He + n + 17.6 \, MeV \qquad\qquad \text{Equation 37}$$

Only in this combination of reactants is the effective radius (expression for the probability of fusion) large enough to ignite at technically feasible temperatures. The situation is outlined in Figure 25.

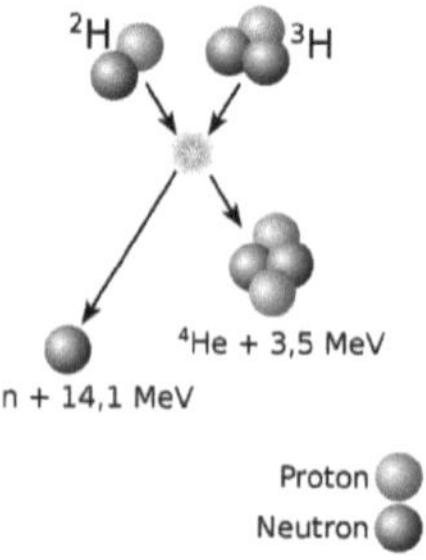

Figure 25   In the reactive collision of a deuterium nucleus with tritium, a helium nucleus and a fast neutron are produced.

Source: https://de.wikipedia.org/wiki/Kernfusionsreaktor

The energy resulting from Equation 37 is distributed as kinematic energy in a ratio of [1 : 4] between the particles α (nuclei^4 He) and the free neutrons formed. The energy of the particles α is further distributed through collisions in the plasma and contributes to its additional heating. If the nuclear reaction rate is high enough (number of reactions per time interval), this energy can be enough to maintain the plasma temperature without additional external heating: the plasma then "ignites" and "burns" on its own. This occurs when the product

Particle density x Temperature x Time constant (determined by the inevitable heat loss),

exceeds a certain minimum value. This is called the energy inclusion time, according to *Lawson*'s criterion[31] .

---

[31]   https://pt.wikipedia.org/wiki/Crit%C3%A9rio_de_Lawson (Tip: often the query is more complete when you search for the subject on wikipedia in German or English, for these pages to be translated into Portuguese by *Google* translator)

However, this point does not need to be reached for a reactor to produce energy. Even at slightly lower temperatures and constant additional heating, sufficient fusion reactions take place. The additional heating also offers a welcome opportunity (in addition to refueling) to control the reaction rate, i.e. the reactor's power output. The plasma state achieved must be maintained permanently, replenishing new fuel according to consumption and removing the resulting helium - the result of fusion. The released neutrons leave the plasma; their kinetic energy, four-fifths of the fusion energy, is available for use.

A few grams of the deuterium / tritium = [1 : 1] mixture are trapped inside the reactor with several cubic meters of content, where it is heated by a high-frequency electromagnetic field, above 100 million degrees. The plasma that forms is directed by a very strong magnet of up to 12 Tesla, so that the plasma does not touch the reactor walls. On the other hand, the fast neutrons do penetrate the reactor walls, as they are not deflected by the electric/magnetic fields. There, the kinetic energy of the neutrons is converted into heat, which can be used to generate electricity via a steam turbine. A second function of the blanket, which consists mainly of lithium, is to produce tritium, which is needed for further fusion (see creation reaction, according to Equation 29).

By increasing the temperature or density, the power produced by the fusion reactions increases. However, it is not possible to reach very high temperatures, as the loss of energy from the plasma due to transport processes also increases with temperature. The desired reaction rate therefore remains constant at constant temperature and density.

Plasma heating methods

Various methods have been developed to heat plasma to over 100 million Kelvin. All the particles in the plasma move at very high speeds depending on their temperature (deuterium nuclei at 100 million Kelvin have an average speed of around 1000 km/s). The heating power increases the temperature and compensates for losses, caused mainly by turbulent and neoclassical transport (caused by collisions between particles) and by braking radiation (*"Bremsstrahlung"*).

Some of the following heating methods can also be used to influence the temperature and therefore also the current distribution in the plasma, which is important for its dimensional stability:

- Electrical heating: Plasma is an electrical conductor and can be heated by means of an induced electric current. Plasma is the secondary coil of a transformer. However, the conductivity of the plasma increases with increasing temperature, so that from around 20-30 million Kelvin or 2 keV the electrical resistance is no longer sufficient to heat the plasma more strongly. In the tokamak, the current is continuously increased through the central solenoid for electrical heating.

- Neutral particle injection: When fast neutral atoms are injected into the plasma (neutral beam injection, NBI for short), the kinetic energy of these atoms - which are immediately ionized in the plasma - is transferred to the plasma via impacts, causing it to heat up.
- Electromagnetic waves: Microwaves can excite the ions and electrons in the plasma at their resonance frequencies (orbital frequency in the helix describing the particle in the magnetic field) and thus transfer energy to the plasma. These heating methods are called "Ion Cyclotron Resonance Heating" (ICRH), "Electron Cyclotron Resonance Heating" (ECRH) and "Lower Hybrid Resonance Heating" (LHRH).
- Magnetic compression: The plasma can be heated like a gas by rapid (adiabatic) compression. An additional advantage of this method is that the plasma density increases at the same time. Only magnetic fields generated by magnetic coils with variable current strength are suitable for compressing plasma.

Establishing the magnetic field

The magnetic field must hold the plasma together against its pressure, so that it does not touch the vessel wall. Both magnetic confinement concepts, *Tokamak* and *Stellarator* (p. 137), use a toroidal and twisted magnetic field. *Tokamaks* create the twisting of the field by inducing an electric current in the plasma, *Stellarators* achieve this through a special and complicated shape of their magnetic coils (more detailed explanation of magnetic confinement and the need to twist the field lines in fusion using magnetic confinement).

Special, localized deformations of the field remove unwanted ions, i.e. the fusion product helium and any impurities, from the plasma (see "diverter" below).

The magnetic field is generated by large coils. The intensity of the current in the coils determines the intensity of the magnetic field and therefore the possible size of the plasma, the density of the particles and the pressure. In a reactor (or in experiments where the plasma is enclosed for a long period of time), the coils must be superconducting: In normally conductive coils, the current flow produces heat due to the electrical resistance that must be overcome. Such coils could no longer be cooled effectively over a long period of operation, which would cause the temperature to rise and the coil to be destroyed. Superconducting coils, on the other hand, have no resistance, which is why the electricity in them does not produce any heat that must be dissipated.

Occurrence and acquisition of fusion fuel

While deuterium is present in the Earth's water in almost inexhaustible quantities ($2.5 - 10^{13}$ t), tritium can only be produced in the necessary quantities through "reproduction" by lithium-6 in the plant itself:

$$\ce{^{6}_{3}Li} \; + \; + \; \ce{^{1}_{0}n} \; \rightarrow \; \ce{^{3}_{1}H} \; + \; \ce{^{4}_{2}He} \qquad\qquad \text{Equation 29}$$

The earth's reserves of lithium are estimated at more than 29 million tons. Only the isotope[6] Li, which occurs naturally in the proportion of 7.5%, is used to produce tritium. From this proportional supply of around 2 million tons of[6] Li, around 1 million tons of tritium can theoretically be extracted using the formula above. In practice, enriched lithium with a lithium-6 content of 30 to 60% should be used. The technically usable lithium deposit is mathematically sufficient to cover humanity's energy needs for thousands of years.

Tritium is radioactive with a half-life of 12.32 years. However, it only emits low-energy β radiation and no associated γ radiation. In the radioactivity inventory of a fusion reactor that has been in operation for some time, tritium will represent a relatively small contribution.

The tritium needed to start up fusion reactors could easily be obtained from conventional nuclear fission reactors. In particular, tritium is produced as a by-product in heavy water reactors (e.g. *CANDU*) in an amount of around 1 kg per 5 GW· year of electricity generated. However, these currently available sources alone would not provide enough tritium for the continuous operation of fusion power plants, resulting in the need to create tritium in the fusion reactor itself. A fusion power plant with 1 GW of electrical output would require around 225 kg of tritium annually.

Tritium creation and neutron multiplication

An economical production of the required quantities of tritium would be possible by producing it from lithium-6 in the fusion reactor itself, as described above, using the free neutrons that are emitted anyway. To do this, the plasma is surrounded by a breeding coat, the **inner mantle of the torus**.

Nuclear fusion provides exactly one neutron per tritium atom consumed. In principle, a new tritium atom could be created from this. However, this is not possible without losses because the blanket cannot capture 100% of the neutrons. Some of the neutrons that hit the blanket are inevitably absorbed into heavier atomic nuclei in the metal alloy (Fe, W, Co, Ni), others escape the system. There are other inevitable losses when the incubated tritium is transferred to the fusion plasma, as well as due to its radioactive decay. So how can tritium production in the fusion reactor be increased?

To supply the plasma with enough tritium, the neutrons in the blanket must be increased by around 30 to 50%. The nuclear processes (n, 2n) carried out on beryllium or lead are used for this purpose (compare p. 84). Commercial fusion reactors must therefore be designed in such a way that a slight overproduction of tritium is possible. The tritium reproduction rate can also be adjusted by the degree of enrichment of the isotope ^{6} Li in the blanket.

Refueling

During the plasma burn time, the fuel must be replenished according to its consumption. Firing pellets made from a frozen mixture of deuterium and tritium into the container has proved to be a suitable technique. Such pellets with a mass of, for example, 1 mg are driven at a speed of around 1000 m/s using a centrifuge or pneumatically with a type of gas gun. This reloading method also makes it possible to specifically influence the spatial density distribution of the plasma by choosing the injection point and the pellet speed. With more or less reloading, the fusion rate can also be controlled; stopping the reloading interrupts the fusion reactions.

Removing helium and impurities

The reaction product, ^{4} He, as well as the nuclei that are inevitably released from the wall material, act as impurities; they must be constantly removed from the plasma. Since they have higher charge numbers than the hydrogen isotopes, this is achieved with magnetic deflection. Specially developed **diverters** are used for this purpose. They consist of deflector plates mounted on the edge of the torus, onto which the unwanted ions from the plasma are directed by means of an auxiliary magnetic field. There they cool down and thus capture electrons again, i.e. they become neutral atoms. These are no longer influenced by the magnetic field and can be removed by the suction system that maintains the high vacuum.

Use of released energy

Of the energy yield of the nuclear reaction, 17.6 MeV per individual reaction, four-fifths, or 14.1 MeV, occurs as the kinetic energy of the released neutron. These neutrons are not influenced by the magnetic field and reach the blanket, where they first release their energy as usable heat through collisions and are then used to produce tritium atoms, as explained above. As in any conventional power station, the thermal energy can generate steam through heat exchangers, which in turn drive steam turbines with attached electricity generators.

Reactor material requirements

The useful energy of the deuterium-tritium reactor comes in the form of high-energy neutrons (14.1 MeV). The neutrons hit the side of the blanket facing the plasma with a high flux density, around 10^{14} s^{-1} cm^{-2} - in addition to exposure to thermal radiation. This inevitably leads to significant radiation damage to the material (for comparison: even in the middle of the core of a typical pressurized water fission reactor, the neutron flux density is about ten times lower, and most of it is "thermal" neutrons).

Radiation damage is highly dependent on the neutron energy. This is why the charge on the wall is often given as the product of the neutron flux density and the neutron energy, i.e. as the power area density in MW m^{-2} (megawatts per square meter). At an energy of 14.1 MeV, 10^{14} neutrons-s^{-1} -cm^{-2} correspond to approximately 2.2 MW m^{-2} . This is the load of the neutron wall proposed in a design for the *DEMO* reactor blanket (p. 137). The blanket should have a service life of 20,000 operating hours, i.e. around 2.3 years. The displacement damage accumulated in this way - which mainly causes embrittlement - is around 50 dpa (displacements per atom) in the steel. In addition, the material is damaged by swelling because the nuclear processes (n, p) and (n, α) in the metal structure produce gases, hydrogen and helium respectively. Helium in the metal is also detrimental to weldability. In order for parts and pipe connections made of steel to be welded again after replacement, a helium concentration below 1 appm ("parts of atom per million", i.e. one He atom per 1 million metal atoms) is required.

In addition, radioactive nuclides are formed in materials through activation. In order to produce the smallest possible quantities, which must also have the shortest possible half-lives, only materials made of certain elements can be used. In today's common structural materials, such as chromium-nickel austenitic stainless steels, neutron activation produces large quantities of the relatively long-lived^{60} Co, which has a high emission of γ rays. Therefore, these steels will not be usable in future power plant reactors.

The main requirements are low activation materials that have sufficient resistance to neutron irradiation and must meet all the requirements for their specific task, such as mechanical stability, weldability, amagnetism or vacuum tightness. It is also currently assumed that the innermost casing must be replaced periodically, as no material can withstand the high neutron flux of a commercial reactor for years. Due to the radiation from the activated parts, repair and maintenance work would have to be carried out remotely, using robots. The aim is to ensure that the majority of activated system components only have to be stored in a controlled manner for around 100 years after the end of their service life, until recycling is possible; a smaller part should be stored for approximately 500 years. Final storage would therefore not be necessary. Current developments focus on nickel-free ferritic-martensitic steels, but alloys based on vanadium and ceramic silicon carbide (SiC) are also being investigated. With *ASDEX Upgrade,* it was discovered that tungsten is also suitable for the front surfaces of plasma-facing blanket modules and diversion plates.

Fusion reactor radioactivity calculations
A spatially detailed calculation of activation in a *DEMO* reactor (p. 137) was presented in 2002 by the Karlsruhe Research Center. A fusion power of 2,200 MW was assumed for the reactor.

Its blanket consists of 77 t of lithium orthosilicate $Li_4 SiO_4$ (enriched in 40% of[6] Li) as the breeding material, 306 t of beryllium metal as the neutron multiplier and 1150 t of "Eurofer" steel (main components 89% iron, 9% chromium and 1.1% tungsten) as the structural material. For all materials, not only the ideal nominal composition was taken into account, but also typical natural impurities, including, for example, a proportion of 0.01% uranium in beryllium. The activity was calculated at the end of uninterrupted operation at full load of 20,000 hours, the useful life of the parts until they are replaced. The radiation dose rate γ on the material surface of a solid component was considered the determining factor for the subsequent handling of the activated parts. It was assumed that reprocessing for new reactor parts would be possible at less than 10 mSv h^{-1} , using remote control technology, and at less than 10 $\square$Sv h^{-1} using direct handling. It turns out that all the materials - lithium silicate, beryllium and steel - can be reprocessed remotely after a decay time of 50 to 100 years. Depending on its exact composition, it takes up to 500 years for activated steel to decay to the point where it can be handled directly.

The total amount of radioactive material generated during the 30-year lifetime of a fusion plant can be estimated at between 65,000 and 95,000 tons, depending on the project. Despite this greater mass, its activity at Becquerel would be comparable to the dismantling products of a corresponding fission reactor. However, the environmental properties would be significantly less burdensome. Unlike nuclear fission power plants, no large quantities of fission products are left over during electricity production, nor ore waste that produces radioactive radon gas.

The road is long and rocky

The technical requirements for a fusion reaction are complex. Here we'll look at the advances that have only been made thanks to a good deal of persistence on the part of researchers.

World record in the last experiment. The reactors currently in operation are all used exclusively to research the technical fundamentals, while a first power plant could perhaps be built within decades. One of these reactors, *JET* (*Joint European Torus*), has been in Culham, Great Britain, since 1983. After 40 years, the research center, in which several European countries are involved, was decommissioned in the fall of 2023 - but before that it was technically tested to the limit. This is how another world record was achieved in October 2023, as the research association *EuroFUSION* now announced in February 2024. An amount of energy of **69 MJ (= megajoules)** was released by the fusion of hydrogen isotopes.

So far, the energy balance has been worrying: 69 MJ is the energy to heat the water of three well-filled bathtubs from about 20 °C to wellness temperature. However: the amount of fuel consumed: < 0.2 grams!

However, with the total energy used to operate the *JET* reactor, it would be possible to heat more than 100 bathtubs. This means that, at the current level of knowledge, the energy investment still outweighs the benefit, by a factor of 30. In order to build a plant, it would be necessary to be able to reverse this relationship between energy expenditure and gain. This would be possible if the mixture of atomic nuclei and electrons contained in the reactor - **the plasma** - could be "ignited". **Ignition**: This is the point at which the fusion reaction can be sustained because it releases so much energy that the plasma remains millions of degrees hotter and a power plant can also function by producing excess heat.

Plasma ignition in the USA has not proved to be a breakthrough. After all, the first successful plasma ignition was already reported in the USA in November 2022. At the Lawrence Livermore National Laboratory in the USA, a small sample of frozen hydrogen was bombarded with powerful lasers. Under the sudden onset of heat, the atomic nuclei fused and energy was released - in fact, more than the laser fired: 2.05 MJ became 3.15 MJ. However, the balance did not take into account the fact that producing the laser radiation also required energy. When this is taken into account, the balance of Livermore's experiment also becomes negative. In addition, the plasma can only be ignited in this way for a fraction of a second - so the fuel has already been used up. Technical concepts for building a reactor in continuous operation do not yet exist. This approach is therefore suitable for research into ignited plasmas, but not for reactor construction. This is different with *JET* and the *ITER* (*International Thermonuclear Experimental Reactor*) experimental reactor, currently under construction in the south of France.

Sun-based energy generation. In the *JET* nuclear fusion reactor, a very thin gas is heated in a tube 4.2 meters high and 2.5 meters wide closed to form a ring (Figure 24). At high temperatures, the atomic nuclei detach from the electrons and take independent paths. A reddish plasma is created and the shelled atomic nuclei collide so violently that they fuse and energy is released - in the form of heat. A similar process takes place in the Sun. Since it has a very high pressure inside, 15 million degrees is enough to force the atomic nuclei to fuse. This high pressure (we're talking 200 billion atmospheres!) cannot be achieved in fusion reactors on Earth. Instead, the plasma trapped inside them must be heated to 100 - 150 million degrees to reach the threshold of nuclear fusion, while the pressure is only a few atm. The magnetic field must hold the particles together against this pressure. Under these conditions, the kinetic energy of the nuclei is high enough to overcome the electrostatic repulsion and approach ~2.5 femtometers (= 10^{-15} m) - the conditions for a reactive collision.

This superheated plasma (its total mass inside the *JET* is a few grams; particle density: ~10^{20} m^{-3}, which corresponds to a fine to high vacuum) must not touch the reactor walls. Not only because it could cause thermal damage to the blanket, but because the plasma would immediately cool down on contact with the walls and the laboriously initiated fusion reaction would be extinguished. Super-strong magnetic fields therefore prevent the plasma cloud from touching the reactor walls.

Technical challenges in generating nuclear fusion. The technically necessary strong magnetic fields can only be produced economically with superconductors - materials that conduct electric current without any resistance, but which require complex cooling. A magnet made of conventional materials will heat up, increase its resistivity and ultimately toast under the high flow of electricity.

The so-called *Tokamak* has been the preferred construction method for fusion reactors until now. The major technical challenge: in order for the plasma cloud in the *Tokamak to* remain compact, an electric current must flow within it. This plasma current is excited and directed by a magnetic field. To build up this magnetic field, an increasing electric current must flow around the reactor vessel. As the current cannot increase to infinity, the *Tokamak* must be switched off and cooled down occasionally. These cycles put enormous strain on the machine.

The alternative concept to the *Tokamak* is the **Stellarator**. It works without the plasma current. However, the calculation of the magnetic fields required in it is so complicated that for a long time it could not be done. But it has become possible thanks to high-performance computers and the *Stellarator*, which many thought was on the fringes, is once again a concept with a future. In the future, the technical elements of the *Tokamak* will probably be installed in the *Stellarator* and vice versa - the technical differences between the two concepts can be unified.

The first plant will be built no earlier than 2050. *JET*'s successor is *ITER* (International Thermonuclear Experimental Reactor), a fusion reactor that is currently being built in the south of France. 35 countries are involved in this project. Development of the plant began in 2007. After many delays, *ITER* is scheduled to go into operation in 2025. Tests will be carried out there, with hopes of achieving a plasma ignition lasting several minutes; this is expected in the 2030s. If everything works optimally, *ITER will be* able to produce approximately as much energy as the system's operation consumes. But only *ITER*'s successor, called *DEMO* (*DEMOnstration Power Plant*), would be the first experimental fusion *power* plant capable of achieving an exothermic balance and being able to feed 300 to 500 megawatts into the electricity grid. However, *DEMO* will not begin operations until 2050.

For an overview of experimental fusion reactors around the world, as well as other technological challenges than those reported here, we recommend you go to https://de.wikipedia.org/wiki/Kernfusionsreaktor.

Alternative concepts to Tokamak and Stellerator

In addition to the fusion of deuterium and tritium in *Tokamaks* and *Stelleratores*, other concepts have been proposed to generate energy on a large scale through nuclear fusion:

- Combinations of fuels other than deuterium-tritium are, in principle, physically possible (for example, deuterium/deuterium, deuterium/helium-3). They would have the advantage of being easier to obtain or of less exposure to radiation. However, they require a significantly higher temperature and/or density for successful operation.
- The concept of inertial confinement is at the basic research stage. In 2022, LLNL (USA) reported the production of 3.15 MJ of fusion energy

from 2.05 MJ of laser energy that hit the target. However, 300 MJ of electrical energy was needed to generate the laser energy.

- According to most scientists, cold fusion is <u>not</u> a possible alternative. The physical processes of energy release claimed in this context contradict the current state of physics.

Nuclear fusion - our energy of the future?

For every realization of a large-scale fusion reactor, proof of economic viability is the decisive success factor.

In order to compare plants with different generation structures, the *Levelized Cost* **of Electricity (LCOE)** was created, which relates the costs of building and operating a plant to the amount of electricity generated throughout the system's lifetime. The unit is therefore R$ per megawatt-hour.

Costs include capital costs (investment for construction) and fixed and variable operating costs. Variable operating costs include fuel costs and emission rights costs. The LCOE does not include the costs of distributing the electricity generated, nor the costs of storing/withdrawing energy to meet peaks and troughs in demand.

An assessment of the economic viability of fusion reactors compared to competing technologies is highly speculative. Studies are not unanimous, but they predict that the LCOE of nuclear fusion will be greater than or equal to that of renewable energy sources. It is certain that investments will dominate, while operating costs are less relevant. Fusion power plants will be large-scale and very capital-intensive projects (in the order of tens of billions of R$), for the central supply of baseload power.

Critics of nuclear fusion note that even conventional and technically sophisticated nuclear fission power plants can hardly compete with the increasingly cheap generation of electricity from the wind and the sun. Storing electricity is also becoming cheaper and cheaper. This restricts the justification for a large power plant to a few locations within a national economy.

The uncontrolled chain reaction

Uranium and plutonium bombs

Natural uranium is a mixture, as we saw above (p. 139) of two isotopes, $^{235}_{92}U$ e $^{238}_{92}U$ which behave very differently. The latter, which is present in great excess, is a neutron absorber. In the nuclear reactor, where the aim was to control the chain reaction of $^{235}_{92}U$, o $^{238}_{92}U$ was still tolerable. But when the aim is to provoke a neutron avalanche, then[238] U must be removed. The uncontrolled reaction requires the most enriched[235] U possible (p. 122).

Critical mass. In the discussion of the natural uranium reactor[235,238] U it was explained that there is a critical size (p. 115) above which the chain reaction propagates. The same goes for reactors working with enriched[235] U; and ditto for the execution of the uncontrolled chain reaction, in other words, for the atomic bomb. The critical size is reached when the [surface : volume] ratio is large enough. From this size there are enough neutrons to sustain the chain reaction inside the fuel. On page 117 this moment was characterized by the "multiplication factor" $k \geq 1$. In the case of pure[235] U, this size is reached with approximately 50 kg of[235] U - which corresponds to a ball 16.8 cm in diameter. By installing "reflectors" (p. 115) on the surface of this ball, its critical mass can be lowered even further. The reflectors, installed tightly around the[235] U, not only lower its critical size, but also prevent early fragmentation of the fuel ball, leading to subcritical pieces. These are factors that increase the detonation power of a weapon, as we shall see.

Below the critical mass ($k < 1$) a block of[235] U is harmless. Above the critical mass ($k > 1$) and in the absence of moderators, there is an exponential growth of fast neutrons, in fractions of a second. This avalanche of neutrons leads to an exponential increase in the rate at which the cascade of reactions 38 takes place. The situation gets completely out of control and there is a devastating explosion, sketched in Figure 26. This is the working principle of the atomic bomb.

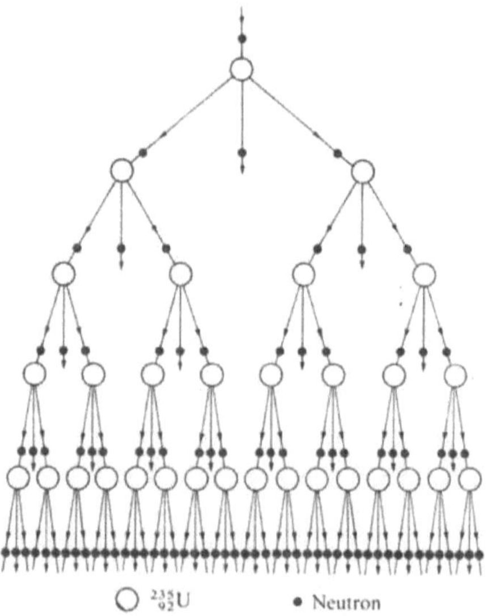

Figure 26 Schematic of the uncontrolled nuclear chain reaction.

At the end of the Second World War, on August 6, 1945, the US dropped a uranium bomb,^{235} U, on the city of Hiroshima (Japan). It caused hundreds of thousands of deaths and injuries (burned and contaminated with radioactive material).

Similar to the^{235} U-bomb, the^{239} Pu also explodes when it reaches above the critical volume. With this fuel material, the critical mass is already reached at 10 to 30 kg, depending on its construction (neutron reflectors). It is easier and cheaper to produce plutonium (p. 118) than pure^{235} U (p. 122). It is therefore more suitable for making an atomic bomb. The second bomb dropped on August 8, 1945 on the city of Nagasaki was a plutonium bomb. It led to Japan's capitulation and marked the end of the Second World War. It remains to be mentioned that just before the two atomic bombs were dropped, on July 16, 1945, the final test was carried out, in the form of a plutonium bomb, near the town of Alamogordo in the state of New Mexico, USA ("*Trinity Experiment*").

How to exceed the critical size, i.e. how to detonate an atomic bomb? There are two possibilities:

> ➤ By means of ordinary explosives, subcritical pieces are thrown together;

> By a strong compression of a subcritical piece that decreases its surface area.

Hydrogen pump

A hydrogen bomb, also known as a thermonuclear bomb, has an even more devastating effect than the atomic bomb. It was first tested by the USA on Eniwetok Island (Marshall Islands in the Pacific Ocean) on November 1, 1952. In this "*Mike Experiment*", what primarily generates the energy is not nuclear fission (of the Equation 38 type), but hydrogen nuclear fusion (p. 106). Its high exothermia is based - just like inside stars - on the transformation of H nuclei into the extraordinarily stable nucleus of ^4_2He. This nucleus can be formed by different paths or isotopes, for example:

$$^1_1H \ + \ ^3_1H \ \rightarrow \ ^4_2He$$

$$^2_1H \ + \ ^2_1H \ \rightarrow \ ^4_2He$$

$$^2_1H \ + \ ^3_1H \ \rightarrow \ ^4_2He \ + \ ^1_0n \qquad\qquad \text{Equations 38}$$

$$^3_1H \ + \ ^3_1H \ \rightarrow \ ^4_2He \ + \ 2\,^1_0n$$

Tritium can be made by bombarding ^6_3L with neutrons, according to Equation 34. This equation can be joined with the penultimate of equations 43 to form the total equation:

$$^6_3Li \ + \ ^2_1H \ \rightarrow \ 2\,^4_2He \ + \ 22{,}4 \text{ MeV} \qquad\qquad \textit{Equation 39}$$

Equation 44 shows the path: the nuclear decomposition of lithium hydride, LiH, which must be initiated at 10^7 - 10^8 degrees (stellar conditions). To momentarily reach a temperature in this range, use is made of the explosion of a uranium or plutonium bomb (see above). Reaction 44 provides less energy than uranium or plutonium bombs, when calculated in kJ per <u>mole</u>. But when related to mass (1 mol LiH = 8 g), it releases 4 times more energy than the (8 g of) $^{235}_{92}U$.

Nova and Supernova

This type of disturbance in the thermonuclear balance can also be observed in space, in the form of a "Nova" or " Supernova ". These terms refer to phenomena in which a star, while emitting a continuous and moderate (or even invisible) light, suddenly becomes very bright for a short time and then flickers out. The Nova with a luminosity of 10^5 times, the Supernova with a light of 10^9 times above normal. It is estimated that in our Milky Way there are 30 - 50 Novae and that only one Supernova occurs every 10^9 stars (rare!). Thus, the Chinese observed a Supernova in 1054, then *Tycho Brahe* (1546 - 1601) documented one in 1572, and finally *Johannes Keppler* (1571 - 1630) in 1604. The increase in luminosity occurs quite suddenly, and then gradually returns to its original luminosity through fluctuations (the star flashing). Today there is a consensus that these phenomena are caused by the star overheating due to growth above a critical size. Above this size, of course, it is no longer possible to dissipate the excess heat being produced inside. Hence an outer layer of this star is being expelled, forming high-speed interstellar matter, its volume is reduced and it is able to emit a greater amount of radiation. Finally, there are two possibilities: either the star returns to its former state of radiative-thermal equilibrium, or it cools down to become a white dwarf (p. 110).

Annex: atomic and cosmic constants

Then the exact values of the quantities used or mentioned in this text.

Table 6 Atomic constants.

Greatness	Symbol / calculation	Value
Unit of atomic mass	$u = 1$ g Z_A^{-1}	$1.6605655 \cdot 10^{-27}$ Kg = 931.5016 MeV c^{-2}
Avogadro's constant	N_A	$6.022045 \cdot 10^{23}$ mol^{-1}
Avogdro number	Z_A	$6.022045 \cdot 10^{23}$
Bohr radius (= minimum atomic radius; radius of H in the ground state)	r_B or a_0	$5.2917706 \cdot 10^{-11}$ m
Boltzmann constant	k_B	$1.380662 \cdot 10^{-23}$ J K^{-1}
Electron: relative mass	M_r (e)	0,000548 58026
Electron: rest mass	m_e	$0.109534 \cdot 10^{-31}$ Kg
Electron: energy equivalent	$E_e = m\, c_e^{\cdot 2}$	0.5110034 MeV
Electron: charge	e	$1.6021892 \cdot 10^{-19}$ C
Electron: magnetic moment	μ_e	$1.001160\ \mu_B$ = $9.284832 \cdot 10^{-24}$ A m^2
Electron: ray	r_e	$< 10^{-19}$ m
Einstein's constant	$E = N\, h_A\cdot$	3.990313 J s$^{\cdot\cdot}$ mol^{-1}
Elemental charge	e	$1.6021892 \cdot 10^{-19}$ C
Faraday constant	$F = N\, e_A\cdot$	96484.56 C$\cdot$ mol^{-1}
Ideal gas constant	$R = N\, k_{A\cdot B}$	8.31441 J K$^{-1\cdot}$ mol^{-1} 0.0820571 atm K$^{-1\cdot}$ mol^{-1}
Constant of gravitation	G	$6.6720 \cdot 10^{-11}$ m$^3\cdot$ Kg s$^{-1\cdot\, -2}$
Speed of light in a vacuum	c	$2,99792458 \cdot 10^{8}$ m s^{-1}
Molar volume of ideal gases	$V_0 = R\, T_0\, /p_0$	22.413831 L$\cdot$ mol^{-1}
Neutron: relative mass	M_r (n)	1,008665012
Neutron: rest mass	m_n	$1.6749543 \cdot 10^{-27}$ Kg
Neutron: energy equivalent	$E_n = m\, c_n^{\cdot 2}$	939.5731 MeV
Neutron: magnetic moment	μ_n	$-1.913148\ \mu_k$ = $-0.966326 \cdot 10^{-26}$ A m^2
Neutron: ray	r_n	$\sim 1.3 \cdot 10^{-15}$ m
Planck's constant	h	$6.626176 \cdot 10^{-34}$ J s$\cdot$

Proton: relative mass	M_r (p)	1,007276470
Proton: rest mass	m_p	$1.6726485 \cdot 10^{-27}$ Kg
Proton: energy equivalent	$E_p = m\, c_p{}^{\cdot 2}$	938.2796 MeV
Proton: magnetic moment	μ_p	2.792763 $\mu_k = 1.410617 \cdot 10^{-26}$ A m$^{\cdot 2}$
Proton: radius	r_p	$\sim 1.3 \cdot 10^{-15}$ m
Atom of $^1_1 H$: relative mass	M_r (H)	1,007825036
Atom of $^1_1 H$: rest mass	m_H	$1.6735596 \cdot 10^{-27}$ Kg
Atom of $^1_1 H$: energy equivalent	$E_H = m\, c_H{}^{\cdot 2}$	938.7906 MeV
Atom of $^1_1 H$: ionization energy	E_I	13.595 eV

Table 7 Cosmic constants.

	Terra	Moon	Sun	Universe
Age	$6 \cdot 10^9$ years	$6 \cdot 10^9$ years	$6 \cdot 10^9$ years	$20 \cdot 10^9$ years
Average speed (orbit)	29,755 Km s^{-1}		250 Km s^{-1}	
Average density	5,514 g cm^{-3} Earth's crust: 2.60 g cm^{-3}	3.342 g cm^{-3}	1,409 g cm^{-3}	$1.227 \cdot 10^{-29}$ g cm^{-3}
Pressure in the center	$3.5 \cdot 10^6$ bar		$200 \cdot 10^9$ bar	
Average distance	of the sun: $149,565 \cdot 10^6$ Km	of the Earth: $3.844 \cdot 10^5$ Km	from the Milky Way center: 33,000 light years [32]	
Pasta	$5,973 \cdot 10^{21}$ t	$7.35 \cdot 10^{19}$ t	$1.989 \cdot 10^{27}$ t [33]	$1.187 \cdot 10^{50}$ t
Surface	$5.0908 \cdot 10^8$ Km^2 [34]	$3.8 \cdot 10^7$ Km^2	$6,072 \cdot 10^{12}$ Km^2	
Average radius	6370 Km	1735 Km	$6.955 \cdot 10^5$ Km	1.3.1023 Km (expanding)
Rotation period	23 h 56 m 3.95 s	365 d 5 h 48 m 46 s	26 d around the Milky Way:	

---

[32] 1 light year = $9.4608 \cdot 10^{12}$ Km

[33] Relativistic mass loss: $4.14 \cdot 10^6$ t s^{-1} ; Radiation power: $3.72 \cdot 10^{26}$ W (= kJ s)$^{\cdot -1}$

[34] Of this, 29.2% is dry land; 70.8% is water.

	around the Sun: 365 d 5 h 48 m 46 s	around the Earth: 27 d 7 h 43 m 11.5 s	$200 \cdot 10^6$ years	
Gravitation		1/6 of the land	28 x the Earth	
Temperature, surface	14.3 °C (average)	150 °C (light side) -150 °C (dark side)	5500 °C	
Temperature, core	20.000 °C		$15 \cdot 10^6$ °C	

Literature and sources used for this book

Most of the research was done on German texts and was freely translated.

Main source: *Holleman, Wiberg*, Lehrbuch der Anorganischen Chemie, 33ª edition, Walter de Gruyter Berlin 1985.

Jörn Müller, Harald Lesch. "Die Entstehung der chemischen Elemente". Chemie In Unserer Zeit **2005**, 39, 100-105.

Contributions to Wikipedia (open source):

https://pt.wikipedia.org/wiki/Modelo_Padrão.

https://en.wikipedia.org/wiki/Nuclear_power_in_space

https://de.wikipedia.org/wiki/Halbwertszeit

https://de.wikipedia.org/wiki/Reinelement

https://de.wikipedia.org/wiki/Kernfusionsreaktor

Other contributions on the internet:

Hermann-Friedrich Wagner " Schneller Brüter";
https://www.weltderphysik.de/gebiet/technik/energie/kernenergie/schneller-brueter/ published on 25/08/2008

I want morebooks!

Buy your books fast and straightforward online - at one of world's fastest growing online book stores! Environmentally sound due to Print-on-Demand technologies.

Buy your books online at
www.morebooks.shop

Kaufen Sie Ihre Bücher schnell und unkompliziert online – auf einer der am schnellsten wachsenden Buchhandelsplattformen weltweit! Dank Print-On-Demand umwelt- und ressourcenschonend produziert.

Bücher schneller online kaufen
www.morebooks.shop

info@omniscriptum.com
www.omniscriptum.com

MIX
Papier aus verantwortungsvollen Quellen
Paper from responsible sources
FSC® C105338

FSC
www.fsc.org

Printed by Books on Demand GmbH, Norderstedt / Germany